AF887280

THE INTERNATIONAL ULTRAVIOLET EXPLORER

CASE STUDY
IN SPACECRAFT DESIGN

PRESENTED BY:

H. R. FREEMAN
PROJECT SYSTEMS ENGINEER

AND

G. W. LONGANECKER
PROJECT MANAGER

GODDARD SPACE FLIGHT CENTER
GREENBELT, MARYLAND 20771

AUGUST 1979

THE INTERNATIONAL ULTRAVIOLET EXPLORER

CASE STUDY
IN SPACECRAFT DESIGN

TABLE OF CONTENTS

THE INTERNATIONAL ULTRAVIOLET EXPLORER

CASE STUDY IN SPACECRAFT DESIGN

1. INTRODUCTION

The International Ultraviolet Explorer, or IUE, was launched successfully on January 26, 1978. Figure 1 is an artists concept of IUE in mission orbit. The satellite was developed to serve as a general facility for observing the ultraviolet spectra of astronomical sources, and during its first year and a half of successful operation, IUE has provided astronomers from many nations well over eight thousand spectral images. Heretofore, the study of the ultraviolet spectrum has been hampered because the Earth's atmosphere prevents most ultraviolet radiation from reaching the Earth. But study of the ultraviolet energy spectra can lead to a better understanding of elements, temperatures, and pressures within stars, and a better understanding of interstellar matter in general. The major element of the IUE experiment package is a 45 cm diameter telescope of f/15 Cessegrain design. The telescope uses an echelle spectrograph for the performance of ultraviolet astronomy in the spectral region between 1150 and 3200 Angstroms and has performed spectroscopy on stars as faint as eighteenth magnitude.

As the name implies, IUE truly was an international undertaking. The satellite and optical instrumentation were provided by the National Aeronautics and Space Administration's (NASA) Goddard Space Flight Center (GSFC), and the vidicon cameras used as detectors in the scientific instrument were provided by the United Kingdom Science Research Council. The European Space Agency (ESA) supplied the solar arrays for the satellite and also constructed a European control center. This international cooperation continues during orbital operations in that two-thirds of the observing time is directed from a control center at the GSFC, and one-third of the time the satellite is operated from the European control center near Madrid Spain, with the latter time divided equally between the United Kingdom and ESA astronomers. The sharing of the observing time roughly reflects the original funding commitments of the three cooperating agencies.

The IUE was conceived in 1969 as a result of recommendations made by the NASA Astronomy Mission Board in their document, A Long-Range Program in Space Astronomy. The document recommended an integrated space astronomy plan for the 1970's including various programs for ultraviolet (UV) astronomy such as the Orbiting Astronomical Observatory (OAO) and Large Space Telescope, subsequently renamed Space Telescope.

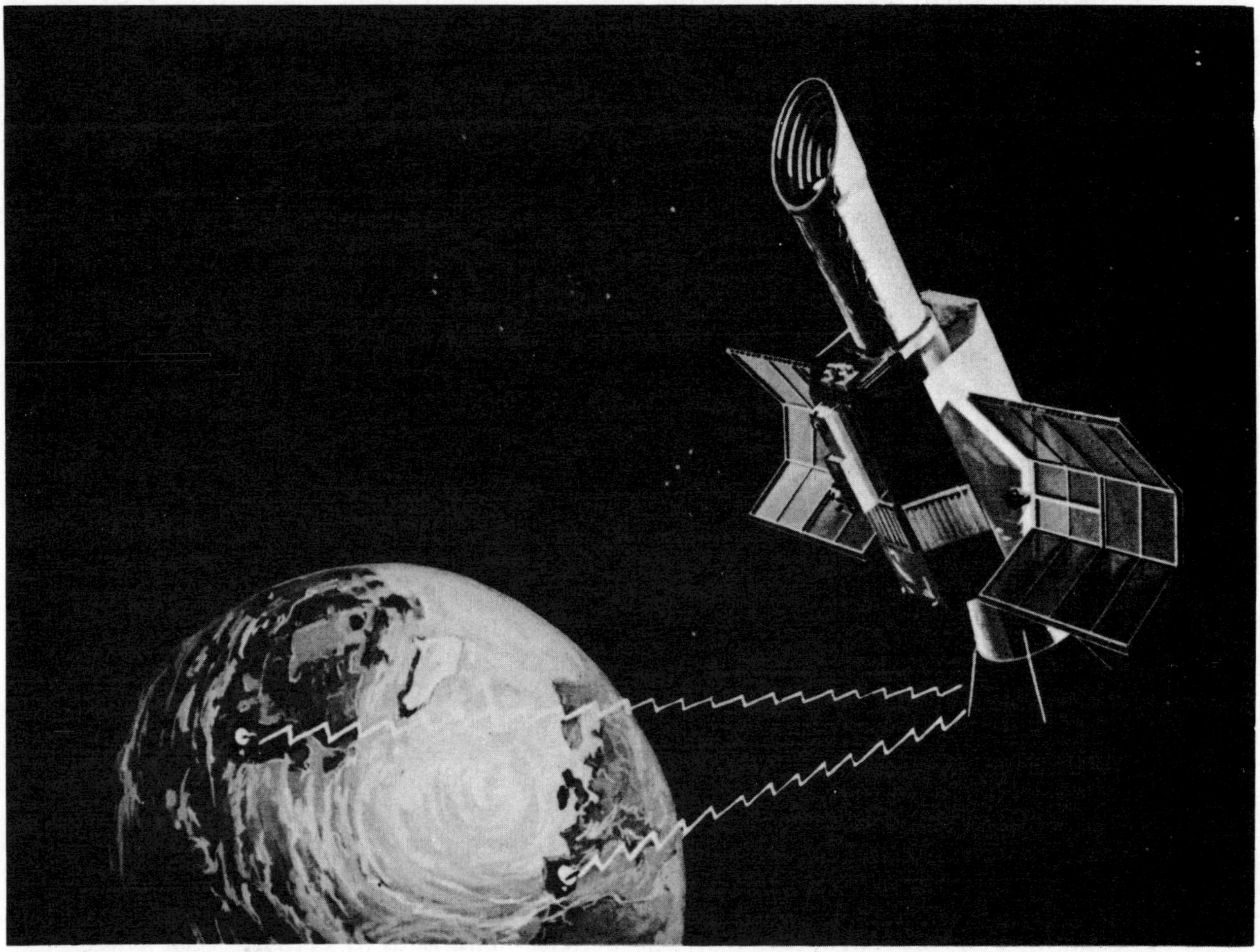

Figure 1. IUE in Mission Orbit

Toward the end of 1969, when further launches of OAO-type spacecraft beyond OAO-C appeared unlikely, the GSFC began an in-house study to see if an Explorer class satellite could meet the needs of the astronomy program after OAO. The study was based on experience in the Small Astronomy Satellite and OAO programs at the GSFC and in the Ultraviolet Astronomy Satellite study conducted for the European Space Agency by the Astrophysics Research Unit of the United Kingdom's Science Research Council. It focused on a system having the following general characteristics:

Three axis attitude control

Instrument with 45 cm UV telescope and echelle spectrograph

Three to 5 year operational lifetime

Delta launch into geosynchronous orbit

Operate as international guest observer facility

Preliminary results of the study were that an Explorer mission of the configuration being studied would have capabilities comparable to the recommendations of the Astronomy Mission Board for follow-on OAO's. Consequently, a formal study effort was initiated in April 1970. Because the system being studied was planned for operation primarily as a guest observer facility, the GSFC established an Astronomy Working Group early in the study to provide guidance on a continuing basis. The original members were chosen for their experience in developing space astronomy instrumentation and because of their interest in using Explorer class satellites for astronomical observations. One of the first actions of the working group was to have a questionnaire broadly distributed to the astronomical community describing the satellite concept and soliciting comments as to the observing programs envisaged and system characteristics desired. More than 170 replies were received from organizations representing half the states and fifteen foreign countries. To help make such broad participation feasible, the system was designed to be operated by observing astronomers in much the same way that ground-based telescopes are operated. As the design of the system progressed, the composition of the working group was changed to reflect this emphasis of operating the satellite as an astronomical observatory.

The study was completed in the fall of 1970 and it established the feasibility of the system concept. Next a system definition phase was initiated to perform additional trade-off investigations to optimize the system and define it sufficiently to begin the flight project phase. Due to agency budget constraints, the systems definition phase was protracted, and flight funding did not begin until fiscal year 1974. Study monies were made available in fiscal years 1972 and 1973 which provided for the design and the start of fabrication of a structural/thermal model of the satellite, the purchase of prototype image tubes for evaluation, and the breadboarding of certain spacecraft subsystems. With flight funding available in July 1973, the goal was to launch in December 1976. However, technical difficulties, particularly with the vidicon camera system, delayed the launch until January 1978.

The IUE is the first astronomical satellite to have been placed into a geosynchronous orbit, and it is the first scientific satellite that allows large numbers of visiting astronomers to make real time observations of ultraviolet spectra. Simplified operational procedures allow the astronomers to joystick the spaceborne telescope about the sky using familiar ground-based observatory techniques. This case study will discuss the technical problems, constraints, trade-offs, and problem solving techniques used during the IUE program.

2. TOTAL PROGRAM OBJECTIVES, REQUIREMENTS AND CONSTRAINTS

The overall IUE Project objectives were:

a. To design, fabricate, test, place into geosynchronous orbit, and operate as an international facility an explorer class spacecraft capable of measuring the ultraviolet spectra of celestial objects;

b. To develop the IUE Observatory (spacecraft and ground system) as an integral system that functionally resembles ground based observatories;

c. To provide a system of maximum utility to astronomers that will extend ultraviolet astronomy research capability to celestial sources of fainter magnitude than previously attainable;

d. To make maximum use of spacecraft instrumentation concepts, designs, and hardware previously developed for other programs, and to promote commonality of hardware between IUE and other current projects;

e. To pursue, to the fullest possible extent, international cooperation in the development and use of the IUE Observatory.

The primary mission objectives were as follows:

a. To obtain high resolution (∿0.1Å) spectra in the ultraviolet region of the spectrum from 1150Å to 3200Å of stars and planets brighter than 7th visual magnitude, for detailed analyses of stellar and planetary atmospheres in order to determine more precisely their physical characteristics;

b. To obtain lower resolution (∿6Å) spectra over the same wavelength range for both stellar and extended objects as faint as 12th magnitude or fainter (15th magnitude desired) as a function of observing time for investigations of peculiar objects such as quasars, Seyfert galaxies, pulsars, X-ray sources, and variability phenomena to shed light on questions of cosmological significances.

Additional objectives were:

a. To evaluate the performance of various subsystems and components such as the spectrograph system (in particular, the image tubes) for potential application to larger astronomical facilities such as the Space Telescope; to evaluate the applicability of an advanced technology 3-axis stabilized Explorer spacecraft for use in other scientific or applications missions.

b. To provide a basis of utility and experience by combining the operations of a geosynchronous observatory and a ground-based realtime observatory in order to maximize international and guest observer participation in this mission and to prepare for potential large telescope missions.

Two other factors were fundamental in determining the design of the scientific instrument and of the spacecraft. First, the instrument should be a spectrograph rather than a spectrometer in order to record simultaneous data over large spectral intervals, and second, the spacecraft should be in continuous contact with the control centers in order to take maximum advantage of the observer's realtime judgment concerning the quality and content of his data.

The requirement for continuous communication between the spacecraft and its dedicated control center dictated that the satellite be in a geosynchronous orbit. In fact, the orbit that was chosen is an ellipse with a 24-hour period, arranged so that the spacecraft is in continuous view of the U.S. ground station and in view of the European station at least 10 hours each day as well. The elliptical orbit was adopted as a result of the spacecraft weight problem that will be discussed later.

The selection of a geosynchronous orbit is fundamentally different from previous orbiting observatories that were only intermittently in contact with ground stations. The latter observatories had to be self-contained, automated systems that acquired data while not under direct control of the guest observer. In the case of the IUE, both spacecraft and scientific instrument control and performance monitoring are exercised continually. The telescope field can be reconstructed on the ground using an image dissector and this image can be displayed to the guest observer via spacecraft telemetry. The observer can identify his target star and direct the course of the observation essentially in realtime. The "Observatory," therefore, is comprised of the ground control center, where the astronomer views his star fields on television monitors, and the optical and electronic instrumentation located at synchronous altitude. The Earth as seen from synchronous altitude subtends an average angle of about seventeen degrees and moves along the ecliptic at the diurnal rate. As a result, the area of sky available to the observer at any given time is much greater than from lower orbits or from the ground and the region of the celestial sphere periodically occulted by the Earth is greatly reduced. Consequently, long exposures and the monitoring of variable phenomena are possible over large portions of the sky without interruption from Earth occultations.

IUE was approved as a one-mission program providing for the design, fabrication, integration, test, launch into synchronous orbit, and postlaunch operation of an Explorer class spacecraft. The program was conducted in the in-house mode; that is, the GSFC acted as the prime contractor. This mode is used periodically as a means of keeping the cadre of scientists, engineers, and technicians current with the state-of-the-art. With this hands-on experience, the GSFC personnel are better able to specify and administer the larger programs using industrial prime contractors. IUE represents the most sophisticated spacecraft program handled to date in-house. While some of the subsystems were procured and delivered as completely tested units, the overall systems design, the design, fabrication and testing of the majority of the spacecraft and instrument subsystems, and the total systems integration, environmental testing and launch support were accomplished in-house.

In order to conserve resources, the complement of hardware was limited to one engineering test unit spacecraft (basically a thermal/structural test unit), one protoflight spacecraft, and one set of critical spare subsystems. The protoflight concept means that the spacecraft was environmentally tested to prototype levels, but flight durations. This obviated the need for a prototype unit. The previously mentioned mission objectives, operational considerations, and overall resource limitations imposed the following mission constraints:

a. Geosynchronous orbit at about 71° west longitude and 29° inclination with the spacecraft within continuous view of the GSFC and within view of the European Ground Observatory for at least 10 hours per day.

b. Spacecraft weight including apogee motor and Delta attach fitting not to exceed the Delta 2914 capability for the mission.

c. Satellite lifetime goal of 3 years with consumables sized for 5 years.

3. GENERAL SPACECRAFT DESCRIPTION

Figures 2 and 3 respectively show the IUE spacecraft in mission-orbit configuration and in an exploded view. The spacecraft's main body is octagonal in shape; its telescope extends from the upper end of the main body; and its fixed solar arrays extend outward from two opposite sides. In mission orbit, the attitude control system maintains spacecraft orientation such that the front of the solar arrays always faces toward the Sun and the thermal louvers face away from the Sun. Thermal louvers, thermal blankets, heat pipes mounted to the underside of the main equipment platform, and heaters within the spacecraft main body and scientific instrument provide thermal control and maintain temperatures within acceptable ranges. The majority of the higher power electronic equipment is located on the main equipment platform within the main spacecraft body and adjacent to the louvers, while the experiment electronics, the attitude control reaction wheels, gyro electronics, and Sun sensor electronics are located on the spacecraft upper equipment platform.

The hydrazine auxiliary propulsion system is located below the main equipment platform. It was required for nutation control, precession, and despin during the transfer orbit operation and is used for Sun acquisition, reaction wheel momentum unloading, station acquisition, and east-west station keeping during the mission orbit. The hydrazine system consists of tanks, plumbing, thruster assemblies, valves, heaters, and supporting structure, and the composite of this hardware forms an integrated, self contained unit. The apogee boost motor which was used to insert the spacecraft from transfer orbit to synchronous orbit is located in the spacecraft lower cone assembly.

The Scientific Instrument, consisting of the telescope and spectrograph, is mounted to the spacecraft structure by means of a strong ring. The strong ring rests on three columns which carry the load to the lower spacecraft structure and these columns are supported laterally by truss members of the main body structure. An inertial reference assembly is mounted directly to the strong ring to simplify alignment and to minimize relative motion between the inertial reference sensor and the scientific instrument. This arrangement permitted alignment of the scientific instrument and the inertial reference assembly as an integral unit and it assures maximum precision with regard to pointing the telescope. The inertial reference assembly is the primary rate and position sensor for the attitude control subsystem and it provides the spacecraft with position stability in the order of a fraction of an arc-second. Fine error sensors and fine Sun sensors afford the inertial reference assembly a drift trim and highly accurate position reference capability. The fine error sensors are in fact two axis star trackers that are mounted within the spectrograph along with the vidicon cameras that are used to store the spectral images.

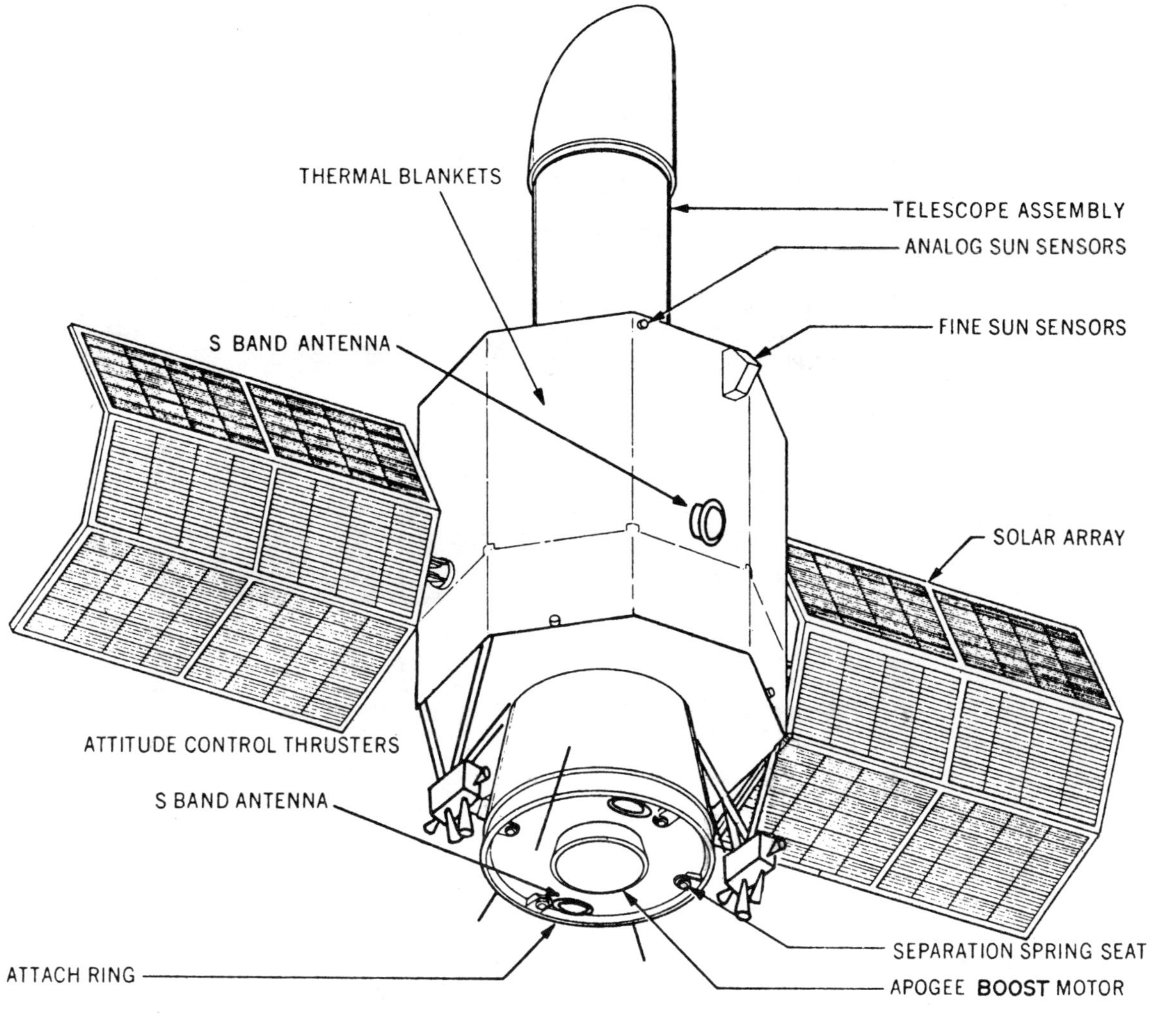

Figure 2. IUE Spacecraft in Mission Orbit Configuration

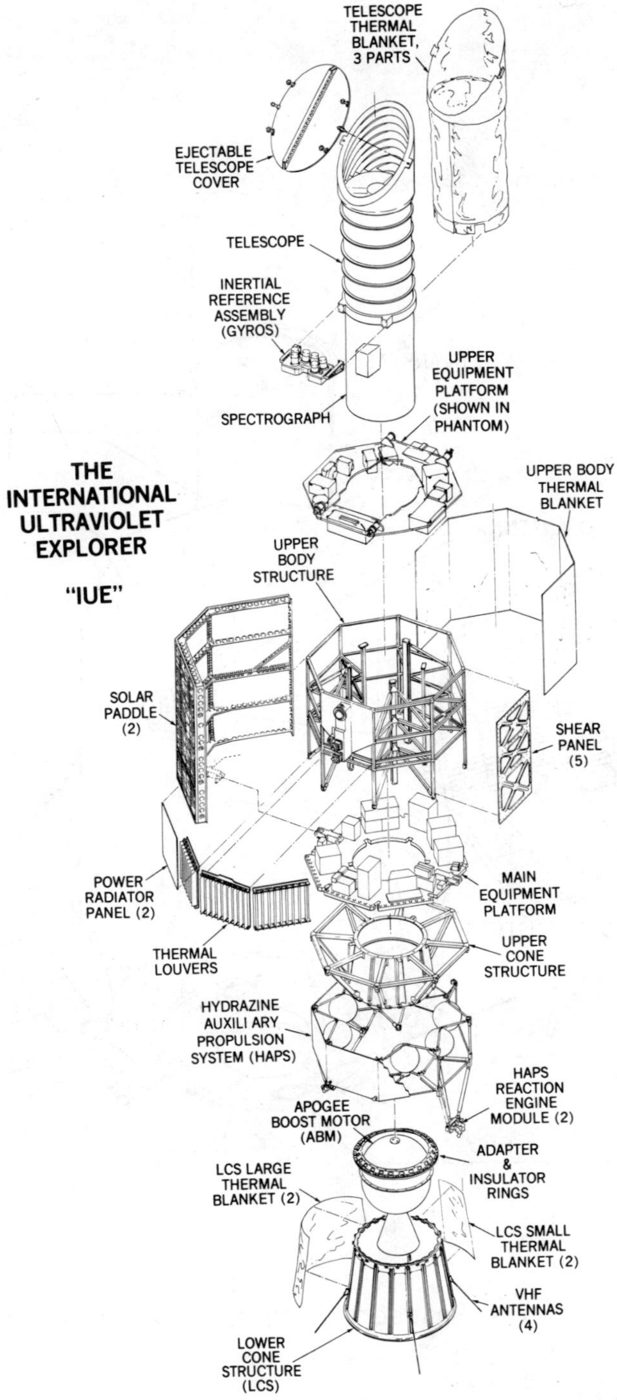

Figure 3. IUE Exploded View

The spacecraft must be capable of pointing with ±1 arc-second accuracy anywhere within the celestial sphere excluding a cone that has a 45 degree half cone angle centered on the spacecraft Sun line. Additionally, it must be possible to reposition the telescope to a new target star that is located up to 60 degrees away with rates up to four degrees per minute per axis, and the newly acquired target must fall within the 10 arc-minute field of the telescope's optical axis. The attitude control system must provide the capability to maintain an imaged target of 1 arc-second within a 3 arc-second diameter aperture hole for extended periods to permit integrated exposures of faint objects using the spectrograph cameras. Storage batteries must be used to support spacecraft and scientific instrument operations during shadow periods, although a thermal design constraint requires that the unpowered spacecraft survive a maximum shadow period without damage. Shadow periods, each lasting approximately 20 days, occur twice a year and the peak shadow period is about 72 minutes in duration. The IUE hardware has a design lifetime of 3 years while the consumables have been sized for 5. Table 1 summarizes the significant spacecraft parameters.

TABLE 1

IUE SPACECRAFT PARAMETERS

CHARACTERISTICS	DESCRIPTION
Spacecraft Weight	312 Kg
Scientific Instrument Weight	122 Kg
Apogee Motor	237 Kg
Launch Vehicle Adapter	29 Kg
Total Launch Weight	700 Kg
Launch Vehicle	Delta 2914
Life	3 - 5 years
Orbit (Mission)	Elliptical Geosynchronous (28.6° inclination)
Power Required (Spacecraft & Experimentation)	210 watts average
Array Capability (Beginning of Life)	424 watts at $\beta = 67.5°$ 238 watts at $\beta = 0°$ and 135°
Batteries (2)	6-AH NiCad Cells (17 each)
Telemetry Bit Rate	1.25 KB/s to 40 KB/s with fixed and reprogrammable formats
Command	PCM/FSK/AM, 800 B/S
Stabilization and Control	Spinning during transfer orbit, 3-axis stabilized with better than 1 arc-second control for mission orbit.

4. SPACECRAFT SUBSYSTEMS

The major subsystems that are required to support the operation of the scientific instrument as well as the spacecraft itself include the power, communications, command and data handling, and stabilization and control subsystems. Figure 4 is an overall system block diagram, and in the discussion that follows, each of the major spacecraft subsystems will be described in detail.

4.1 POWER

The IUE power subsystem is a Direct Energy Transfer (DET) system where the primary source of spacecraft power is the two deployable solar paddles. Solar array power is transferred directly to the spacecraft bus which is regulated to +28.0 volts ± 2 percent. Regulation is accomplished through the use of dump resistors attached to shear panels that are located on the antisun side of the spacecraft. The lack of series elements between the solar array and the spacecraft loads provides for a nearly 100% efficient transfer of array power to the loads. Two 6 ampere-hour nickel-cadmium batteries, operating through a boost regulator, supply power during solar eclipses and during periods when the power demand exceeds the capability of the array.

Specific IUE power requirements involve two distinct phases of the mission; the ascent/transfer orbit phase, and the mission orbit phase. During the transfer orbit period, the solar panels were stowed along the facets of the spacecraft and the spacecraft was spinning at 60 rpm. Accordingly the array output was considerably diminished. The panels were deployed for the mission orbit phase and the spacecraft pointing constraints maintain the Sun within an acceptable view angle relative to the array so that array power is sufficient to support the normal operating loads. The power subsystem was designed to have sufficient battery capacity to support a minimum spacecraft load for a maximum shadow period.

The design of the power subsystem was influenced by the following basic requirements:

a. Standarization of basic subsystem functions for maximum commonality with other spacecraft designs.

b. Use of conventional solar conversion/energy storage system with proven design techniques.

c. Three-year mission design life with a 5-year design goal.

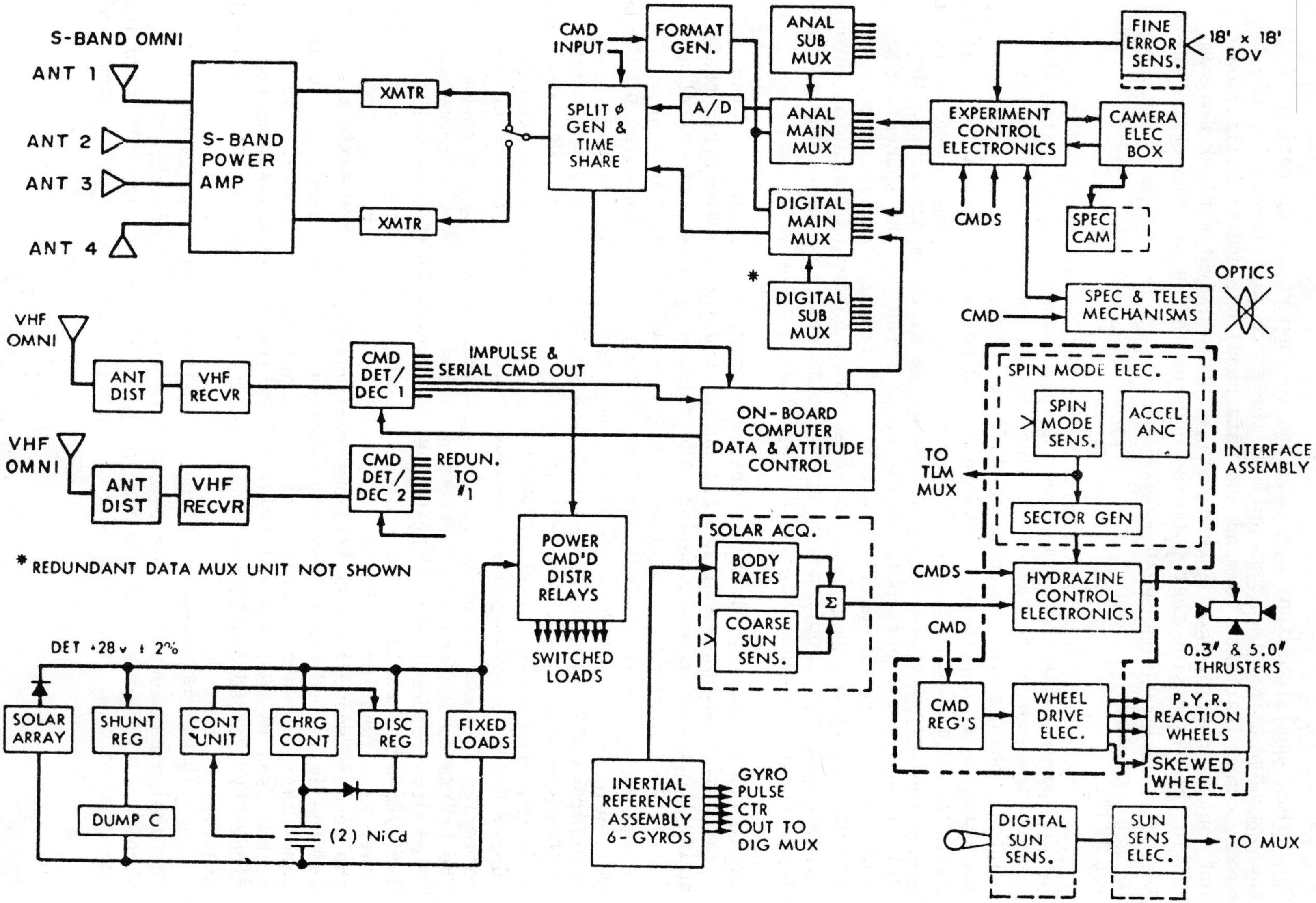

Figure 4. IUE System Block Diagram

d. Need to support mission orbit operations for Sun angles of $\beta = 0°$ through $\beta = 135°$ and for roll angles up to $\pm 15°$ (the Sun is normal to the spacecraft roll or telescope axis for $\beta = 90°$ and looks up the bottom of the spacecraft for $\beta = 0$).

e. Need to support transfer orbit operations for $\beta = 90° \pm 45°$.

f. Redundant units where necessary to assure maximum confidence in achieving design goals.

g. Prohibition of automatic switchover to redundant units because of reliability considerations.

Requirements a. and b. were satisfied by incorporating design techniques and/or component designs that were successfully flown on other satellites, while requirement c. concerned itself with battery life, array degradation, piece part failure rates, and radiation damage. The size and configuration of the array was predicated by requirements d. and e.

With regard to the stowed array power, figure 5 illustrates the instantaneous solar array output as a function of spacecraft rotation angle for a typical β angle and figure 6 shows a plot of average and minimum power for various β angles. This curve clearly shows that if one were to depend solely upon array power for this phase of the mission, the available power would be limited to P_m, the minimum power per revolution. If the battery were used to augment the array generated power, then the power system could support continuous load equivalent to Pa where the shaded area above the line labeled Pa must equal the shaded area below the line Pa multiplied by the factor of inefficiency commensurate with the battery supplied power. Furthermore, the useful continuous power load line must be derated from Pa to Pu whenever there is a transfer orbit shadow. This derating is necessary in order to supply the additional power that would be required to replenish the batteries as a result of continued spacecraft operation in shadow. The estimated IUE transfer orbit and shadow times were 14 hours and 72 minutes respectively. Figure 7 illustrates the mission orbit deployed array while figure 8 shows the deployed solar array available power as a function of β angle for a variety of conditions. It was the need to support scientific observations for $\beta = 0$ through $\beta = 135°$ and the desire to use a simple non-driven array that drove the solar panel design toward the three subpanel configuration shown. The method used to meet the requirements concerning redundancy and the prohibition of automatic switching are discussed as part of the general description that follows.

The design consideration for commonality with other spacecraft significantly influenced the design of the IUE power supply electronics. The concept of a standard power conditioning unit for a DET system was

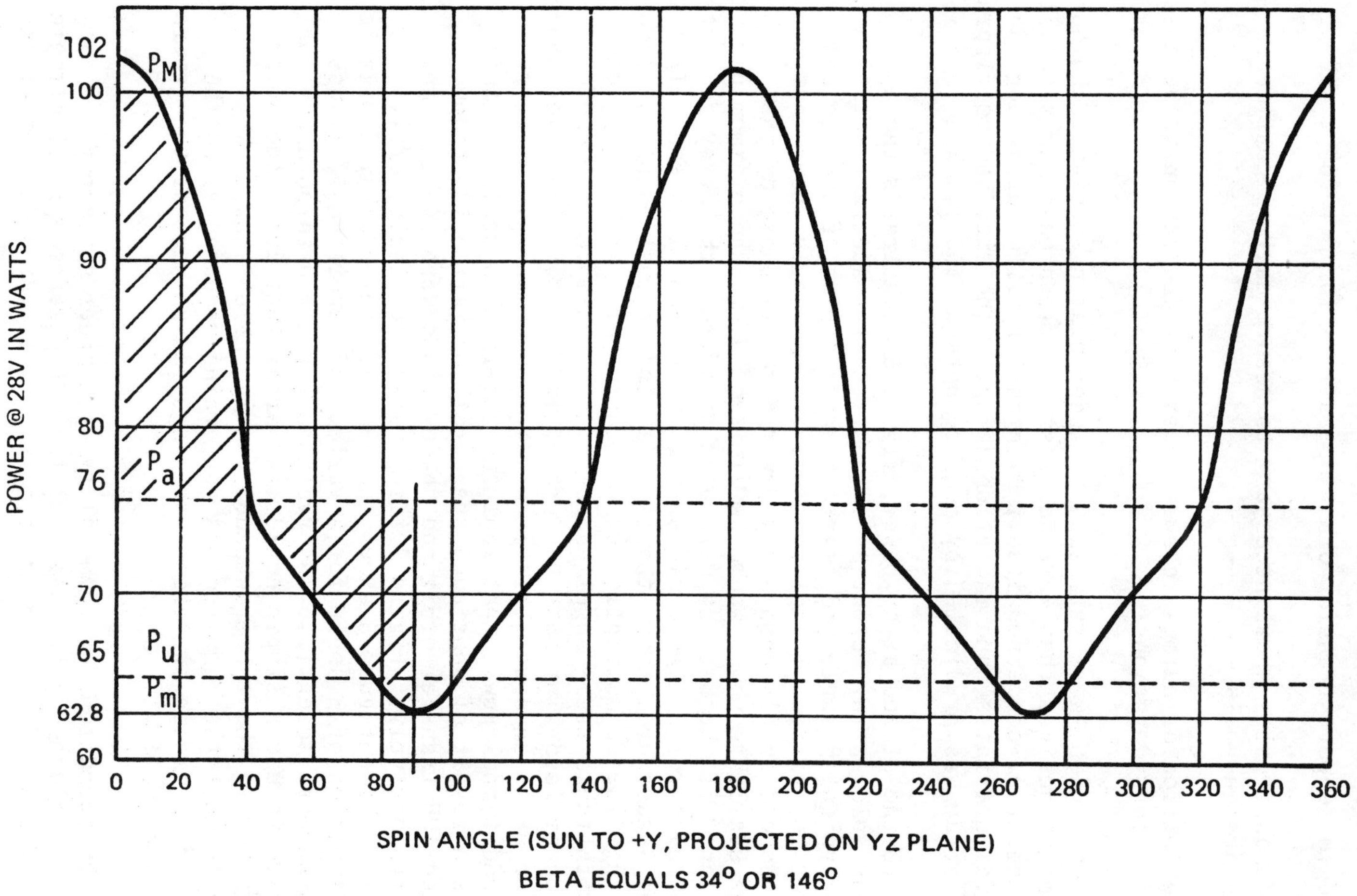

Legend: P_M = Maximum power/revolution
P_a = Power Available
P_u = Useful Power
P_m = Minimum power/resolution

Figure 5. IUE Solar Array Power Output in Stowed Configuration

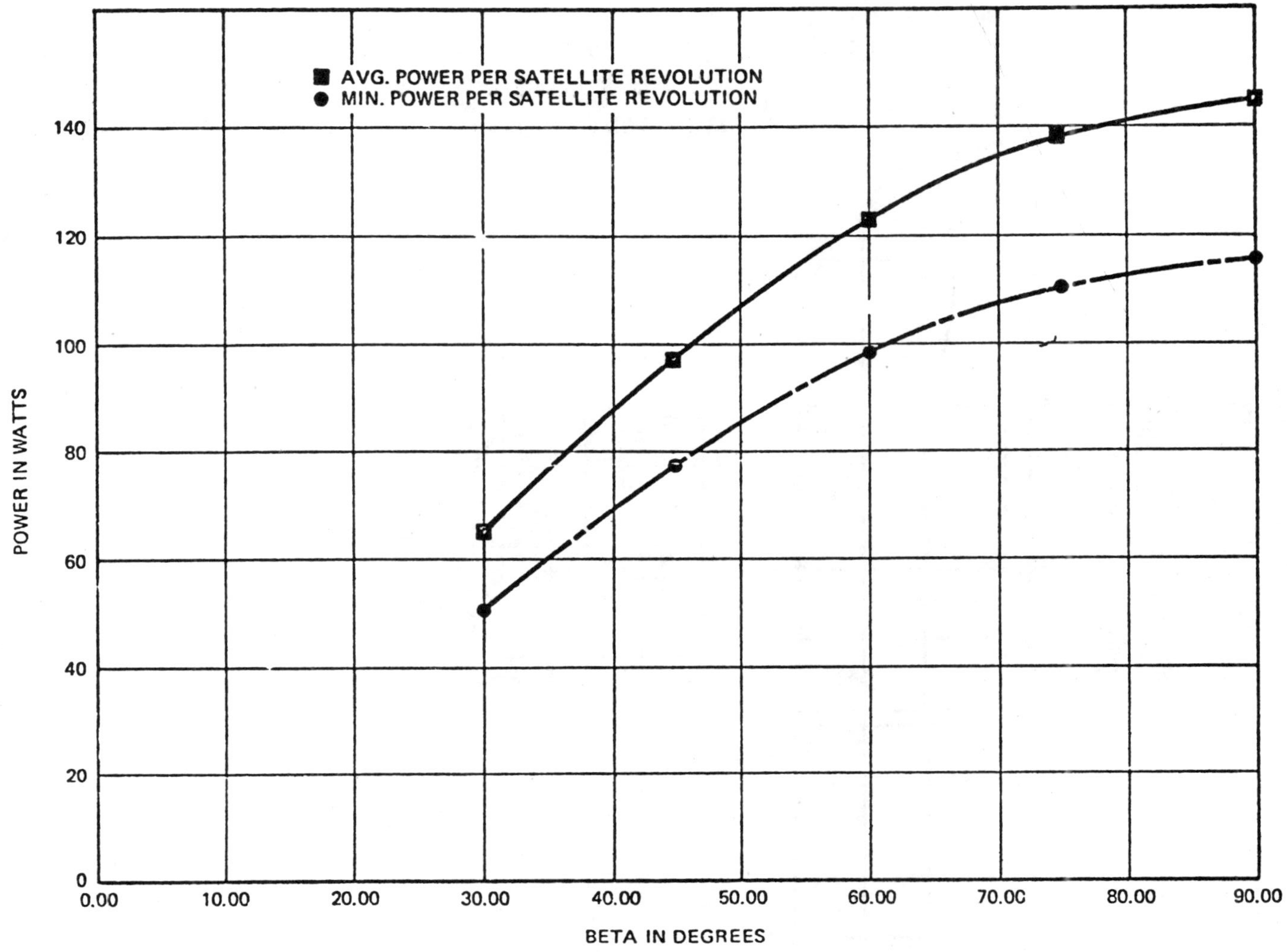

Figure . 6. Power vs Beta for IUE Solar Array in Stowed Configuration

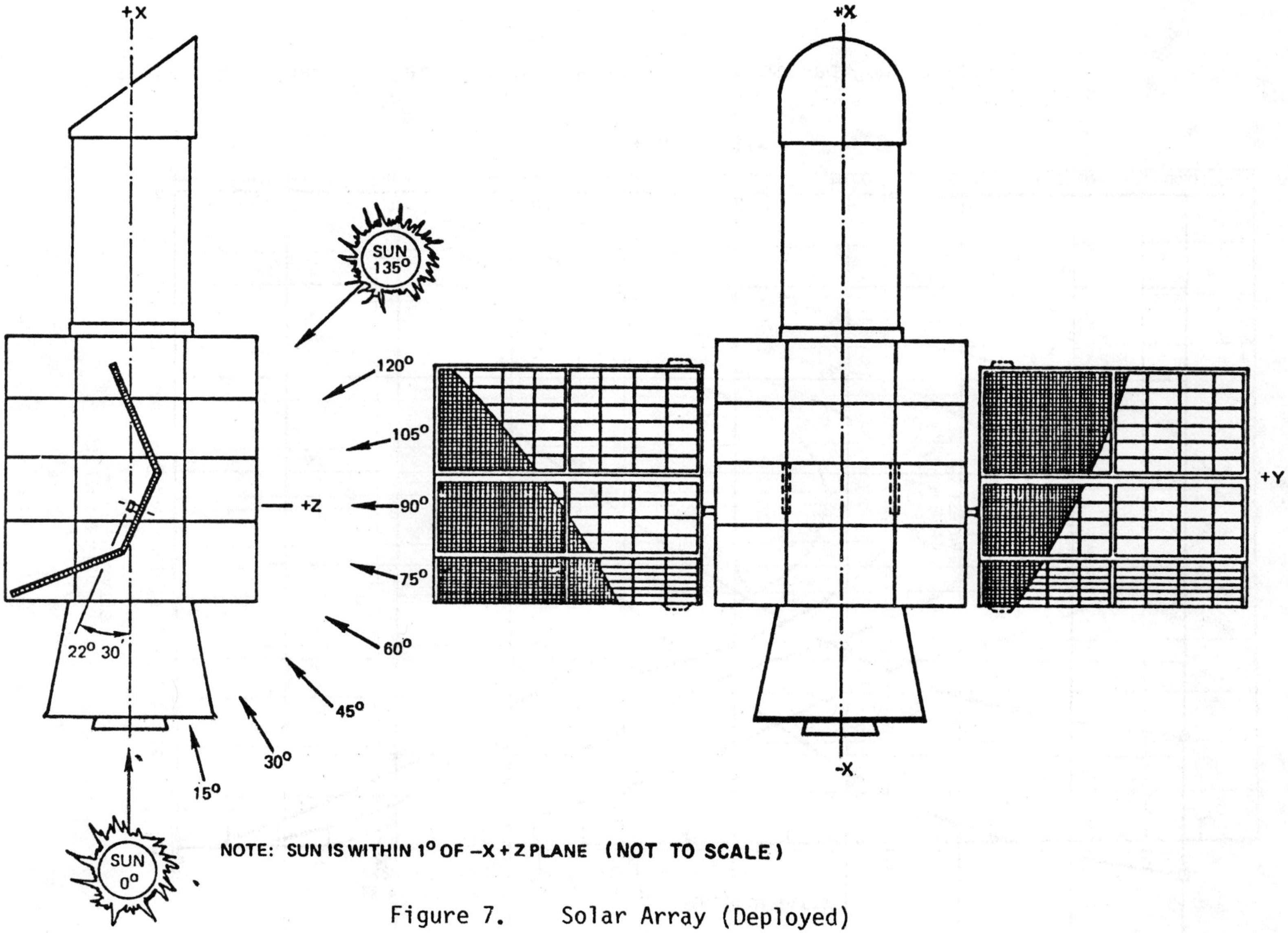

NOTE: SUN IS WITHIN 1° OF −X + Z PLANE (NOT TO SCALE)

Figure 7. Solar Array (Deployed)

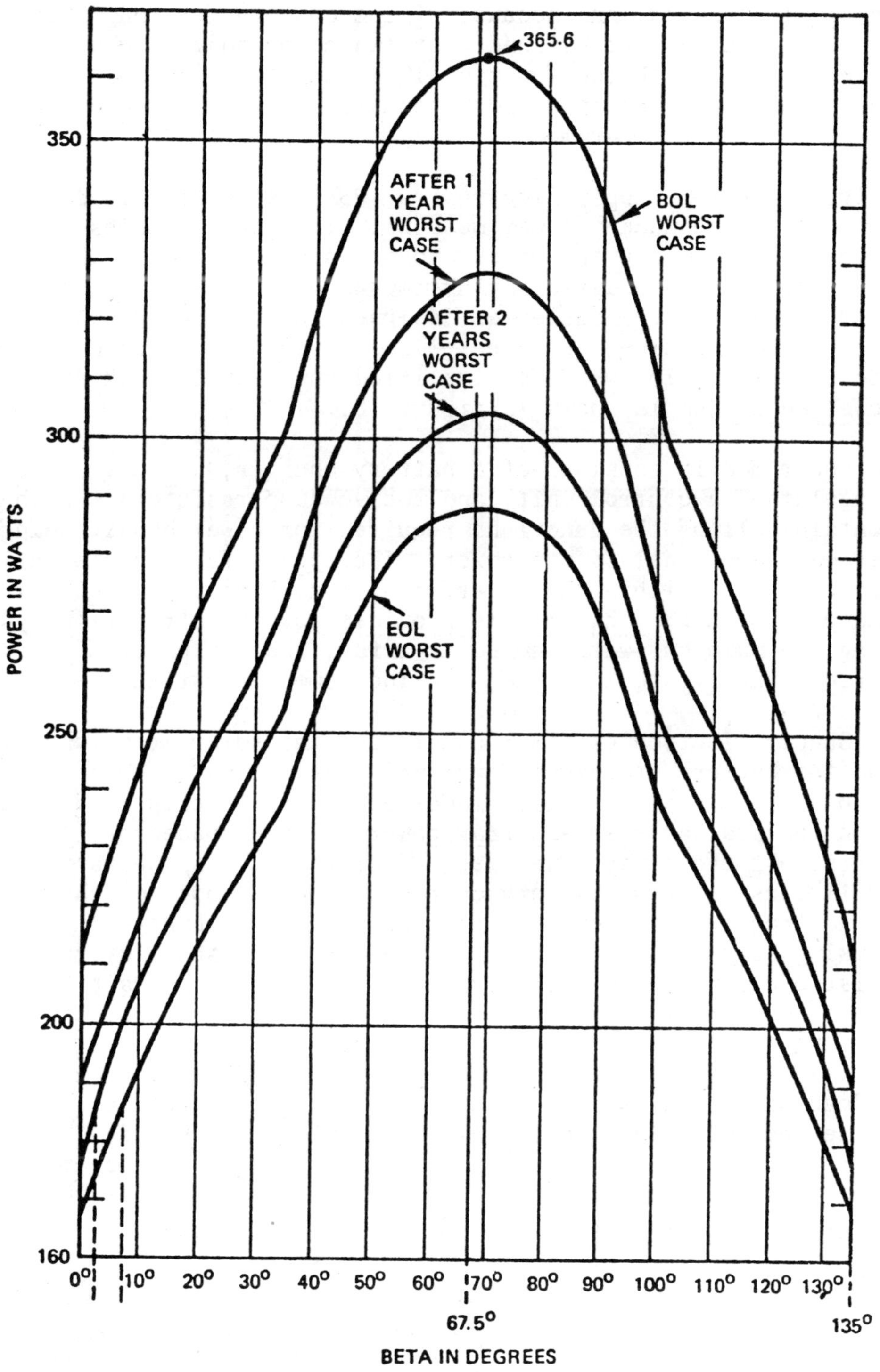

Figure 8. Solar Array Available Power

explored. This study resulted in a modular approach for the power conditioning electronics, and because of the concern for long life and reliability, it was established that two power modules would be operated in parallel with the functional elements of each module being capable of continued operation even after a failure of another function within a particular module.

Accordingly, the Power Supply Electronics (PSE) is modular in design and is composed of redundant power modules. The power modules are operated in unison and are interfaced to one another as well as to the redundant batteries and solar arrays by means of a Mission Adapter Module (MAM). Each power module was designed to sustain 300 watts for indefinite periods of time. Figure 9 illustrates the IUE power supply electronics configuration while a simplified functional block diagram of the power subsystem is shown in figure 10.

The basic power module consists of a battery charger, a shunt driver, a boost regulator, a control unit, and twelve dump resistors. Each module contains all of the functions required for power conditioning and either can handle all of the power. The successful implementation of this concept lies within the interfacing box which is called the Mission Adapter Module. This module provides power distribution by interfacing the main bus with the solar array, the two power modules, the batteries, the pyrotechnic bus, and the command system.

The Power Supply Electronics (PSE) conditions the solar array energy source and the battery storage element and provides a spacecraft regulated bus of 28 v ± 2%. Three modes of operation exist and these depend upon the available solar array power and the spacecraft load requirements. If the array power is greater than the spacecraft load, the PSE will first provide battery charge current to both of the battery chargers and then dump the excess array power to the shunt elements in sequence using both sets of shunts. A typical shunt/shunt driver characteristic curve is shown in figure 11. If the spacecraft load exceeds the solar array output, the difference in power will be supplied by discharging the batteries through the boost regulator. If the spacecraft load is equal to the solar array power, the PSE will operate within a dead-band, and when operated in this region, neither the charger, the shunt nor the boost will be actively controlled. To ensure a proper transition between each of these modes of operation the controller was designed with three distinct bands of operation, and these are illustrated in figure 12.

4.2 COMMUNICATION

The communication subsystem consists of redundant VHF transponders, redundant S-band transmitters, and four power amplifier/antenna systems. This subsystem interfaces with the power subsystem, the telemetry

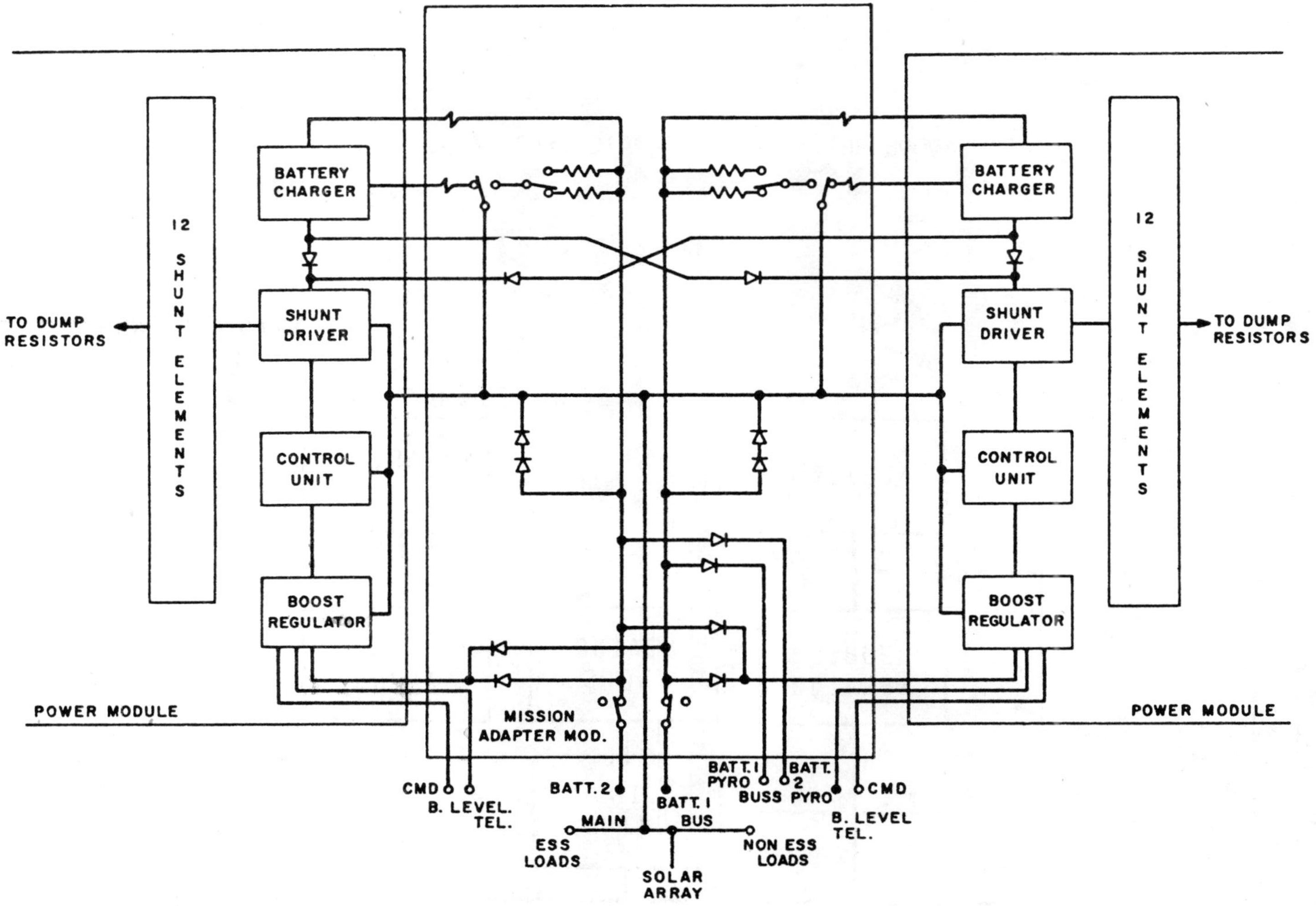

Figure 9. IUE Power Supply Electronics

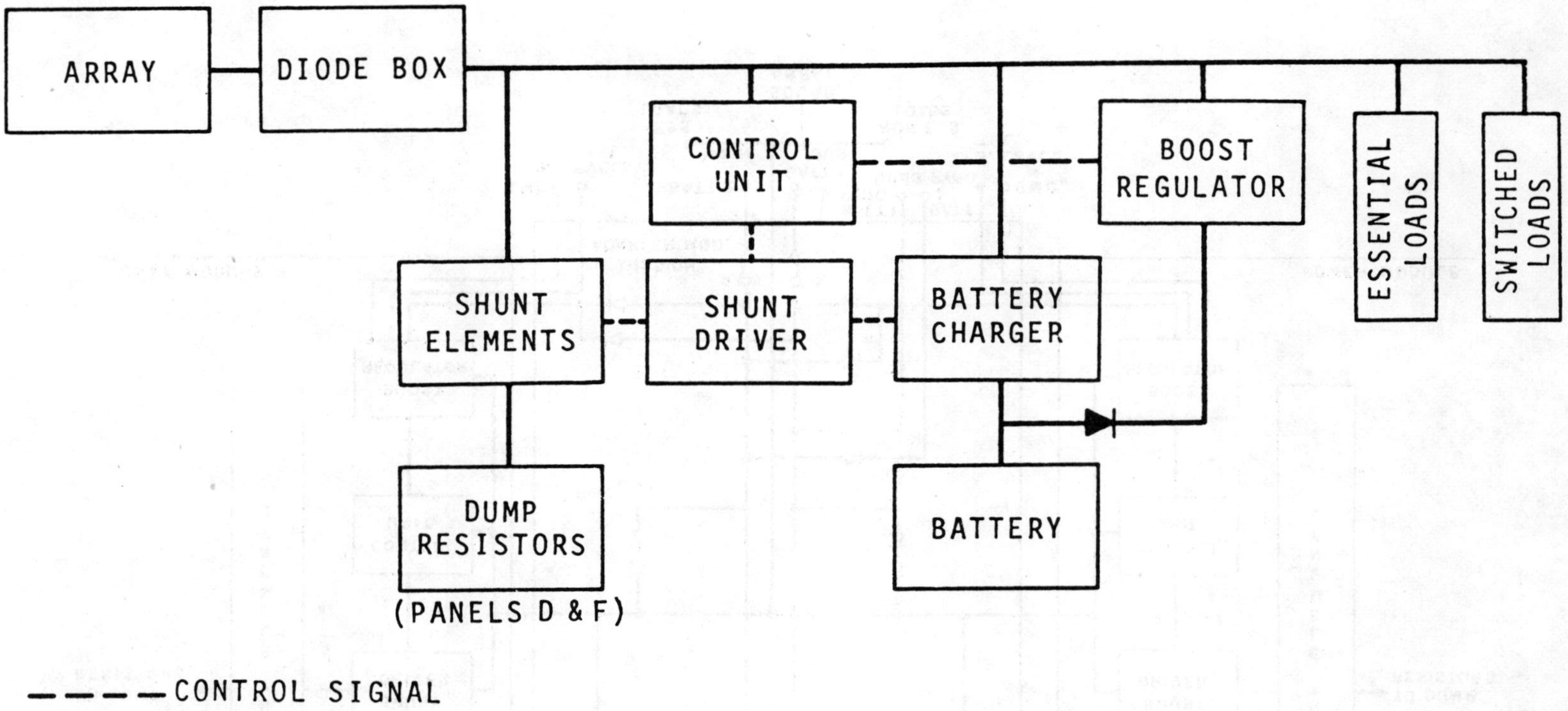

Figure 10. Functional Block Diagram - IUE Power Module

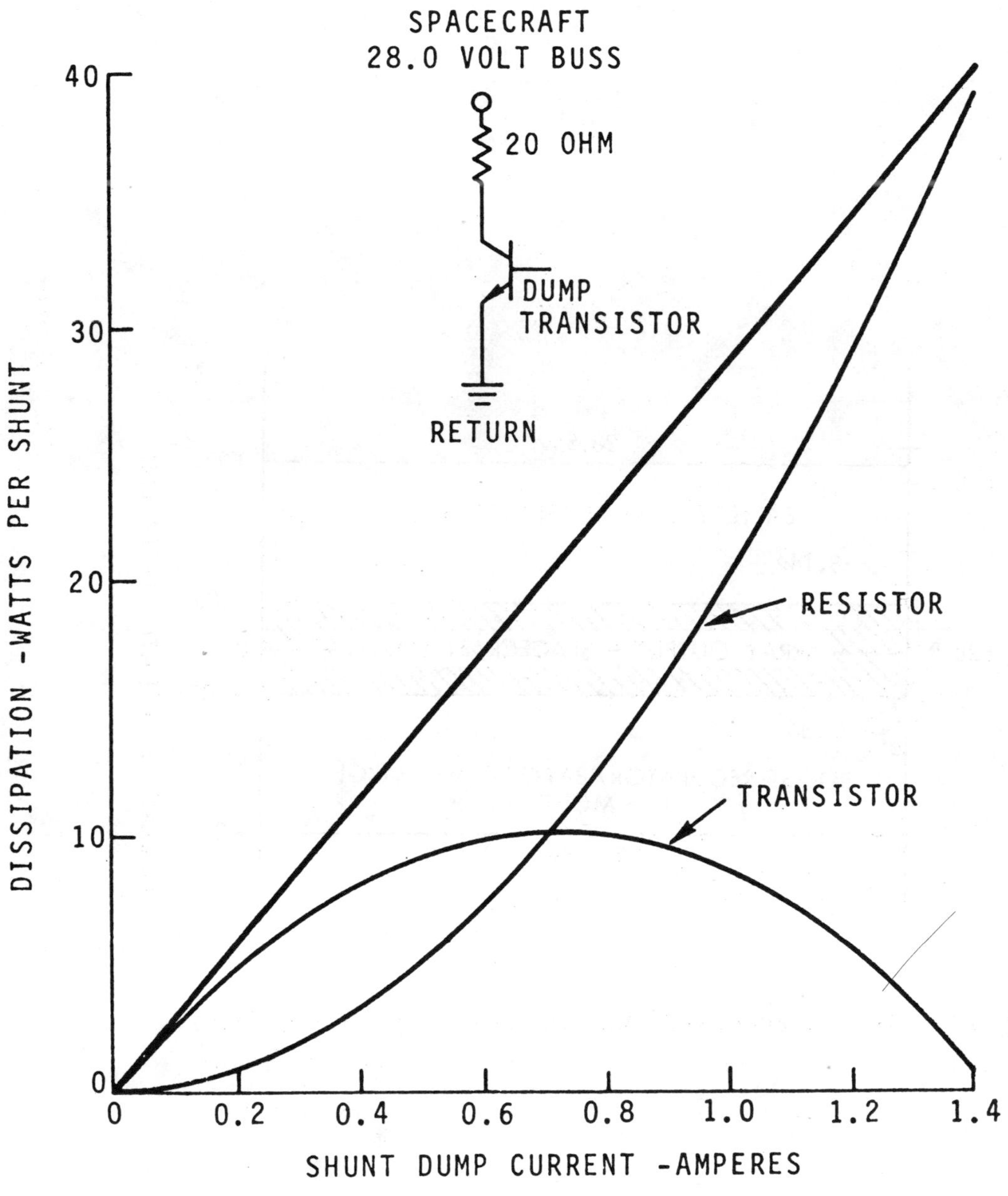

Figure 11. Operation of IUE Power Subsystem Shunt Element

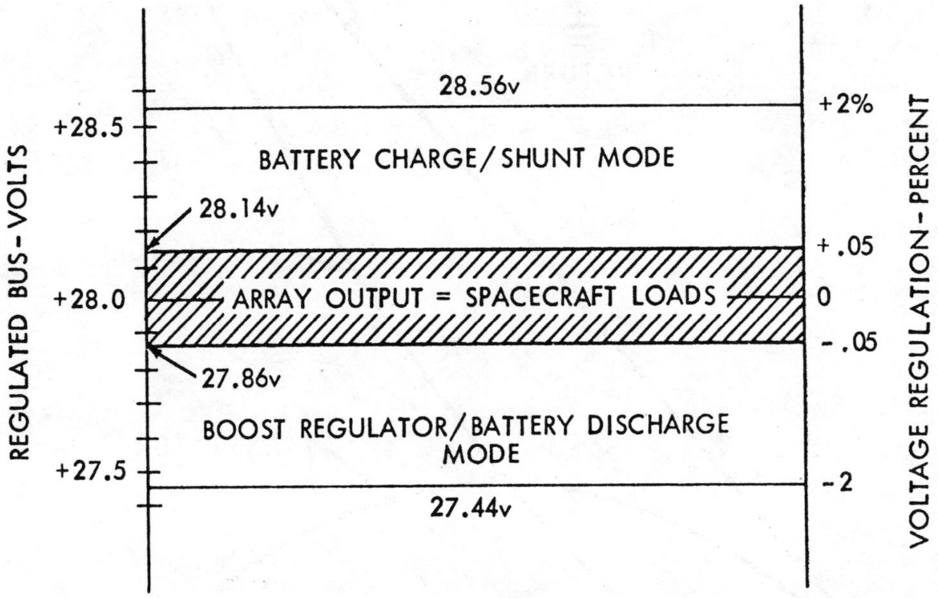

Figure 12. Spacecraft 28.0-Volt Bus Regulation Modes

encoder output, and the command decoder input. Functionally, it provides for redundant methods of transmitting telemetry data to the ground, it provides for the reception of ground generated commands and it provides a range and range rate measurement capability for orbit determination using the Goddard Range and Range Rate System.

The design of the communication subsystem was influenced by the following design requirements.

a. Use of flight proven hardware and design techniques as well as a high degree of reliability.

b. The need to support commanding and the transmission of telemetry from transfer orbit through mission orbit operation.

c. The need to support telemetry rates up to 40 Kb/second for a variety of spacecraft/Earth pointing angles.

d. The need to provide for a method of tracking the spacecraft.

During the pre-mission orbit phase of operation, the spacecraft orientation relative to the Earth was not well defined and the data rate requirement was low. Accordingly, a VHF omnidirectional system was flown to satisfy both the telemetry and commanding requirements for the transfer orbit and acquisition mode operations. Throughout the mission phase, however, the spacecraft/Earth orientation is well known and a directional S-band downlink system was flown to accommodate the high data rates required by the scientific instrument. Selectable S-band antennas provide coverage for all allowable spacecraft/Earth pointing angles. The VHF transponder continues to provide the command up-link capability and additionally it affords the spacecraft a commandable range and range rate mode of operation that provides for the turn-around transmission of the range and range rate signals for metric tracking of the spacecraft using the Goddard ranging system. Figure 13 illustrates the locations of both the VHF and the S-band antennas.

The VHF system consists of two transponders, an antenna distribution system, and a turnstile antenna system. A functional block diagram is presented in figures 14 and 15. As can be seen, each transponder consists of a power and switching module, a 6 watt transmitter, a receiver, and a range and range rate module. The VHF receivers provide the exclusive means for the ground to command the spacecraft, and each receiver is dedicated to one of the two command decoders. Clearly, this is a critical function and commanding is possible even when the transponder is in its ranging mode of operation. Finally, each transmitter can provide telemetry to a ground station from either of

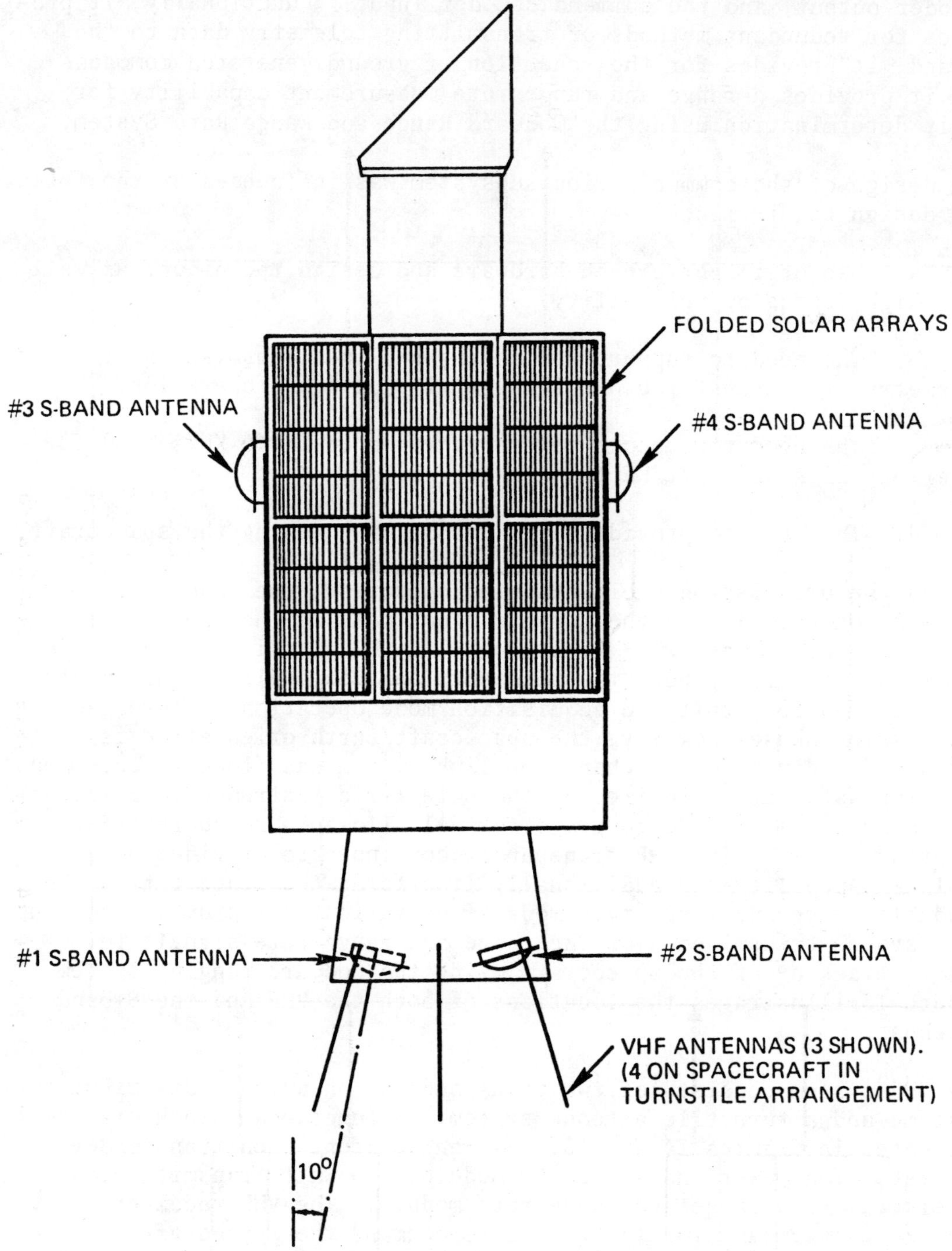

Figure 13. IUE RF System Antenna Locations

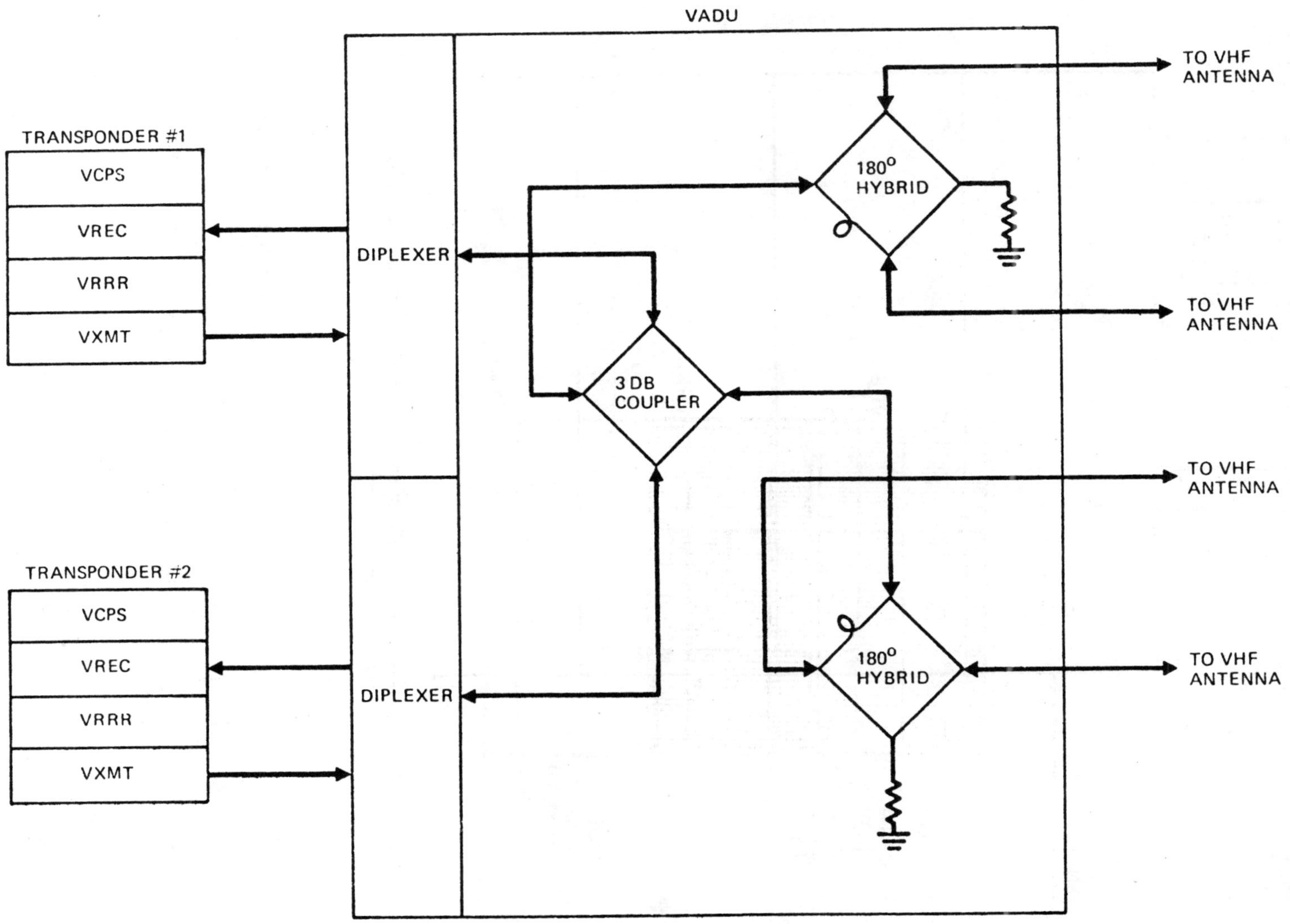

Figure 14 . VHF System Block Diagram

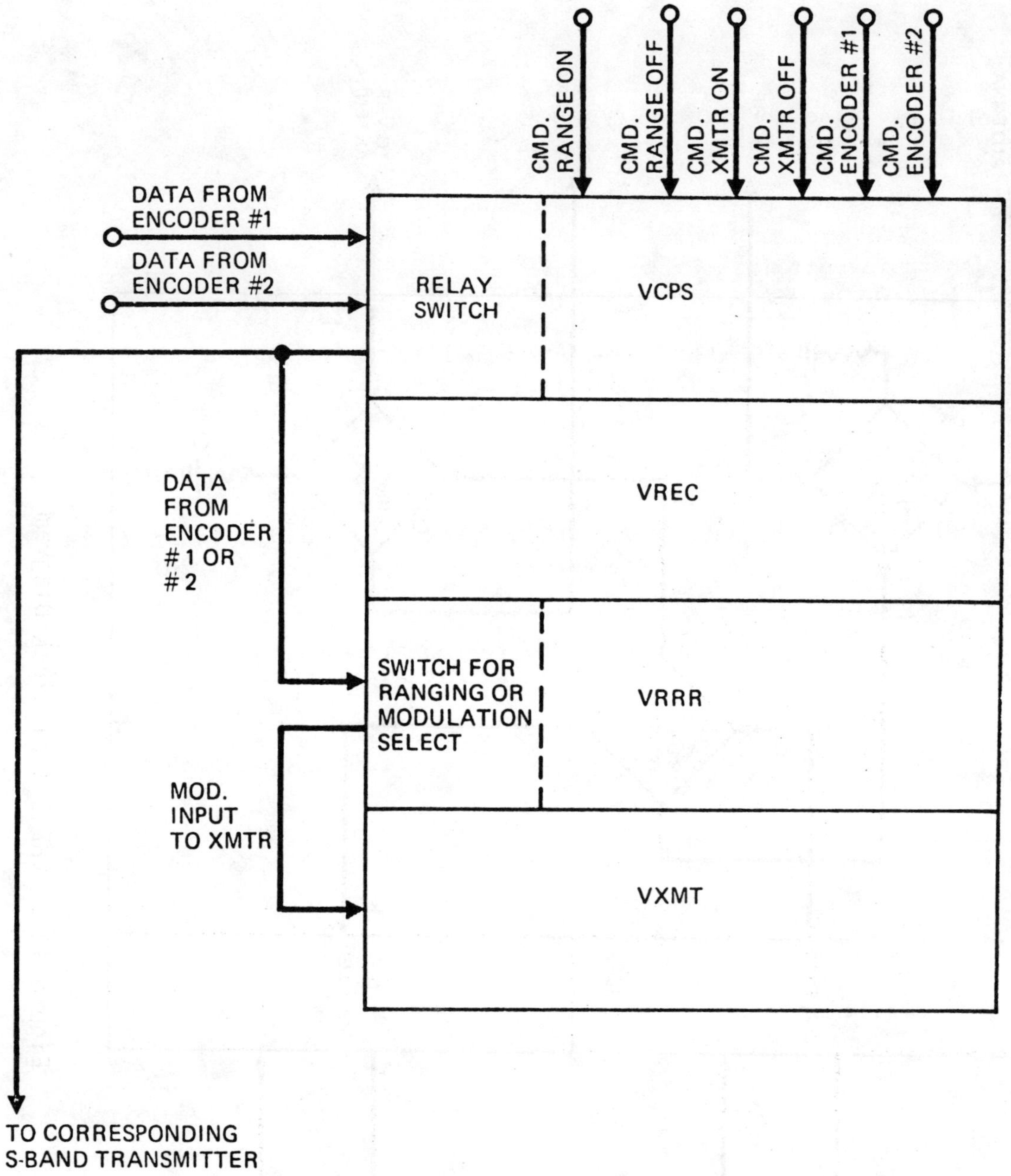

Figure 15. VHF Data Flow Diagram

the redundant data units at a bit rate of 1.25 kilobits per second. This capability can be used at any time during the mission, and accordingly, the VHF link serves as a back-up to the S-band telemetry link. The transponder characteristics are summarized in Table 2.

The S-band system consists of two redundant transmitters and four S-band antennas; each is driven by a dedicated 6 watt power amplifier. Figure 16 illustrates the S-band system block diagram. Both transmitters are connected to each of the power amplifier/antenna combinations but only one transmitter and one power amplifier is active at any one time. The selection of a particular power amplifier is executed via ground command, and this selection is based upon the knowledge of which antenna has the most favorable spacecraft-to-ground look angles for a given spacecraft orientation. Each transmitter can provide telemetry from either of the redundant data units at bit rates of 1.25 to 40 kilobits per second in binary increments. Table 3 summarizes the characteristics of the S-band system.

4.3 COMMAND AND DATA HANDLING

The command and data handling subsystem consists of command decoders, command relays, data multiplexers, and an on-board computer. Functionally, this subsystem accepts and decodes the commands that it receives from either the ground or from the on-board computer, it encodes both spacecraft and scientific instrument telemetry, and it performs all of the attitude control system computations. Additionally, the on-board computer issues commands, monitors the health of critical subsystems, and accurately controls the exposure times of the spectral images. The following requirements influenced the design of the command and data handling subsystem.

a. Use of flight proven hardware and design techniques as well as a high degree of reliability.

b. The need for real time as well as delayed command capability.

c. Ground initialized and data/event initialized commanding.

d. Both serial-digital and discrete command word capability.

e. Ground selectable and reprogrammable data formats, and ground selectable data rates and multiplex ratios.

f. Continual ground visability of scientific and housekeeping data.

g. Autonomy with regard to spacecraft safety.

TABLE 2

CHARACTERISTICS OF THE IUE VHF TRANSPONDER SYSTEM

Receiver Frequency:	148.980 MHz
Transmitter Frequency:	136.860 MHz
Power Output:	6 Watts
Output Modulation:	PM
Receiver Sensitivity:	-106
Antenna Polarization:	Turnstile
Antenna Pattern:	Omnidirection

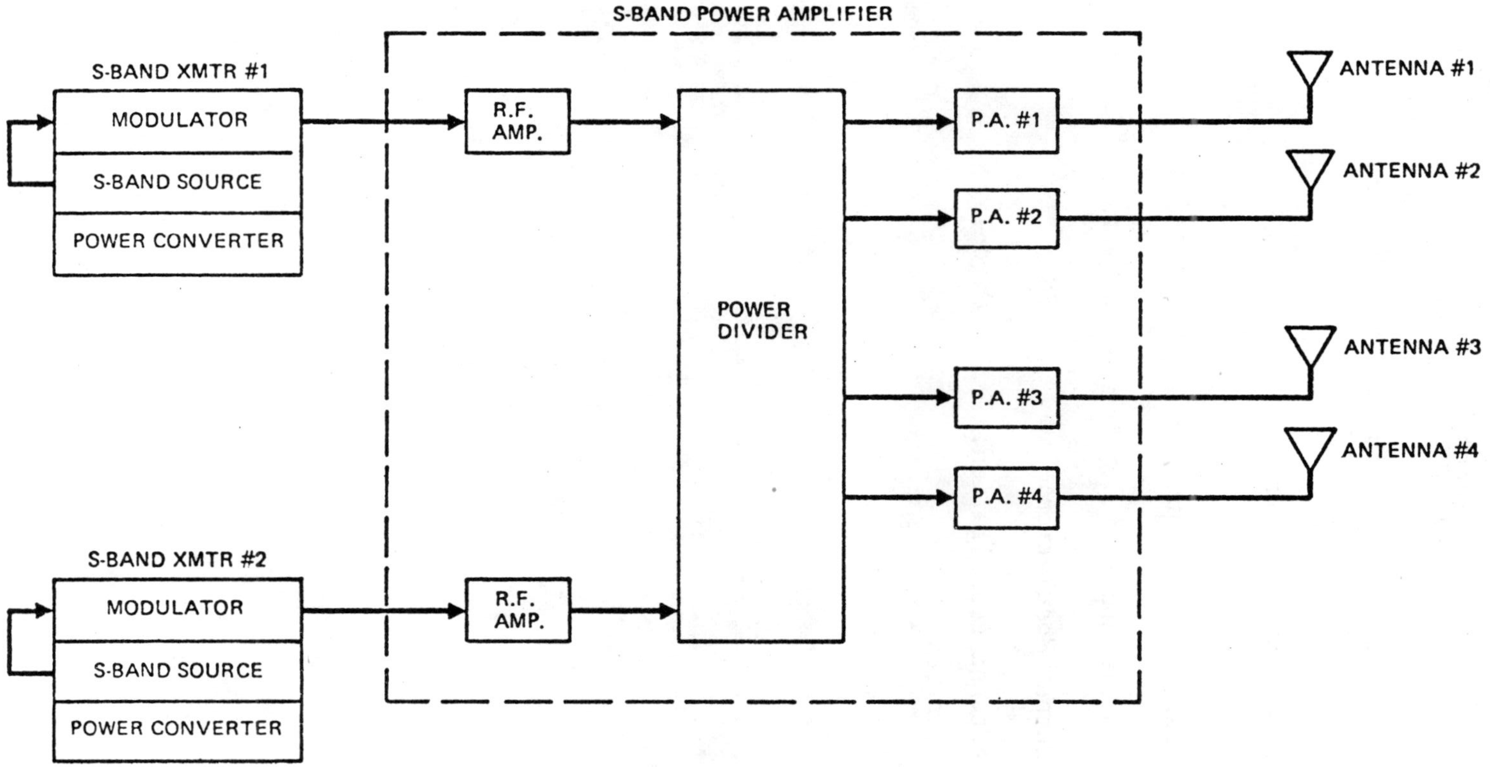

Figure 16. Block Diagram of S-Band System

TABLE 3

CHARACTERISTICS OF THE IUE S-BAND SYSTEM

Transmitter Frequency:	2249.80 MHz
Power Output:	6 Watts
Output Modulation:	PM
Antenna Polarization:	Circular

h. The need for an on-board computer to mechanize the highly complex control algorithms and to perform time critical functions.

4.3.1 Command

Figure 17 illustrates a functional block diagram representation of the command portion of the command and data handling subsystem. The Command System consists of two redundant command decoders and a single command relay unit. Each decoder can process messages from two sources: a VHF receiver analog signal or from digital information generated by the on-board computer. The decoders detect these messages and route each of them to the appropriate spacecraft system or to the scientific instrument. The relay unit provides relay closures and distributes primary power to all spacecraft systems, and although the relay unit itself is non-redundant, all critical relay closures are, in fact, functionally redundant.

A second function provided by the CRU is the pyrotechnic circuitry. This circuitry controls, by serial and discrete commands, the apogee boost motor ignition, the solar array deployment, and the telescope cover release. The pyrotechnic circuitry is completely redundant.

Command conflicts or the need to establish priority between ground generated and on-board computer generated commands are avoided by the use of time-shared control of the command decoder execution logic. Through the use of this logic, alternating time intervals are available to interleave real-time ground generated and on-board computer generated commands. Each redundant decoder is dedicated to one of the VHF receivers and each is individually identifiable by a spacecraft/decoder address.

As a result, each command message to the spacecraft will identify only one decoder. The decoder addressed will accept the command message while the other will indicate that a false address has been received, and this decoder will enter an alarm state. Once in the alarm state, no further data input is recognized by a decoder until the decoder is reset. To reset a decoder simply requires that the subcarrier be interrupted for a fixed time interval.

Ground uplinked and OBC command data are serially transferred, stored, and execution is performed based on timing generated by the decoder time-share logic. Alternate fifteen millisecond intervals are provided for the command execution of ground and OBC generated commands respectively, and because the time share intervals of the Command Decoders are not synchronized, it is important to ensure that the uplinked and OBC commands are addressed to the same decoder. If this condition is violated, the possibility exists that the uplinked and the OBC commands to an individual user could overlap.

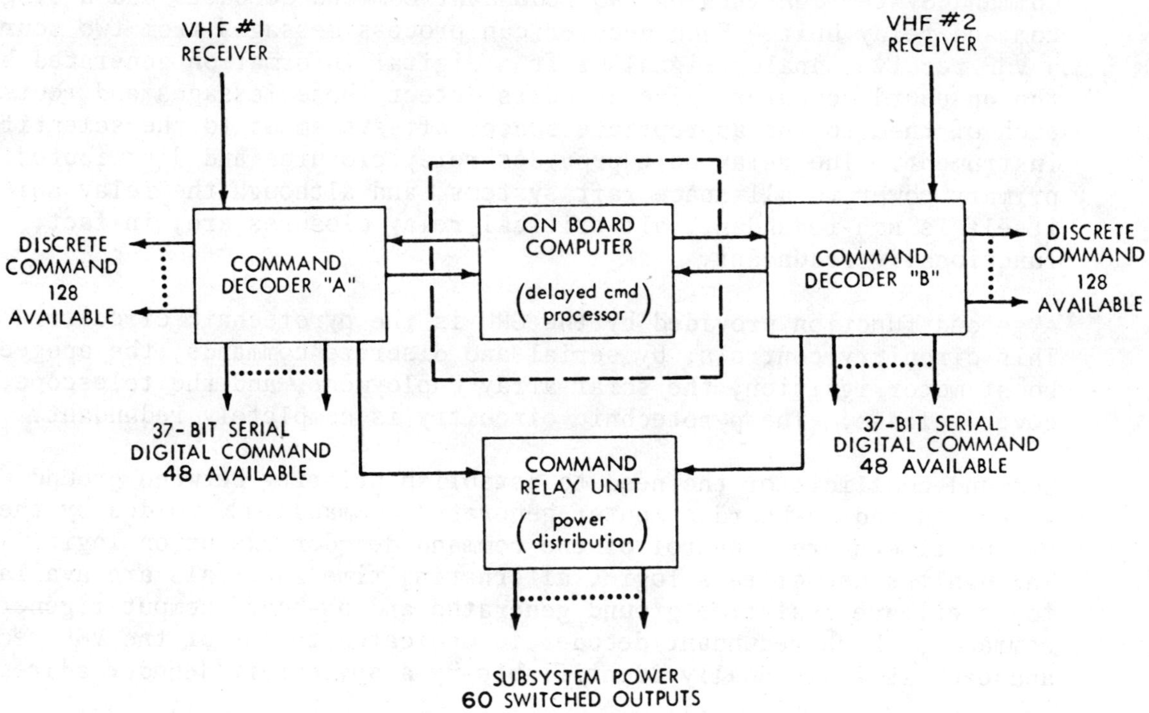

Figure 17. Block Diagram, IUE Command Subsystem

The command decoder issues two types of commands, discrete commands and serial commands, and the command capacity of each decoder is 128 discrete or impulse type commands and 48 serial digital commands. The discrete commands provide a ten volt fifteen millisecond duration pulse while each serial command provides for the transfer of 37 bits of programmable information. The command format consists of 60 bits of information at a bit rate of 800 bps. The bit rate clock is amplitude modulated by ONE and ZERO tones that are frequency shift keyed. The ONE tone frequency is 12 kHz and the ZERO tone frequency is 8 kHz. The command execution time for ground commands is 75 milliseconds, while OBC commands can be executed on 30 ms centers. Each of the serial and discrete commands have been assigned to an individual subsystem and/or black box, and each command can be issued to either command decoder for redundancy.

The command relay unit provides primary power distribution and relay closure functions using 60 commandable latching relays. Additionally, the command relay unit provides pyrotechnic control using redundant arming relays and redundant momentary-closure firing relays. Each of the arming relays is controlled by the command relay unit serial command while the firing relays are controlled using impulse commands. The serial command that is assigned to the CRU controls the state of each of the latching relays and a single command can be issued that would open all relays. This command is issued automatically in the unlikely event that a bus or battery undervoltage or bus overcurrent should occur. Issuance of this command removes all switched loads from the spacecraft bus so that fault isolation can proceed. Figure 18 illustrates a typical relay driver circuit.

The spacecraft command system contains protection against erroneous commands, and when an incorrect command is received, a telemetry flag is set and the carrier must be dropped before additional commands can be accepted by the command decoder. Each decoder contains a command counter that increments each time a valid command is received via one of the VHF command receivers.

4.3.2 Data Handling

A functional block diagram of the data subsystem is shown in figure 19. This subsystem serves as the spacecraft telemetry encoder for all subsystems. As such it provides properly formatted telemetry to the communications subsystem and to the on-board computer on a time shared basis.

The Data Subsystem contains two identical data multiplex units and each multiplex unit consists of one dataplexer, one analog subplexer, and

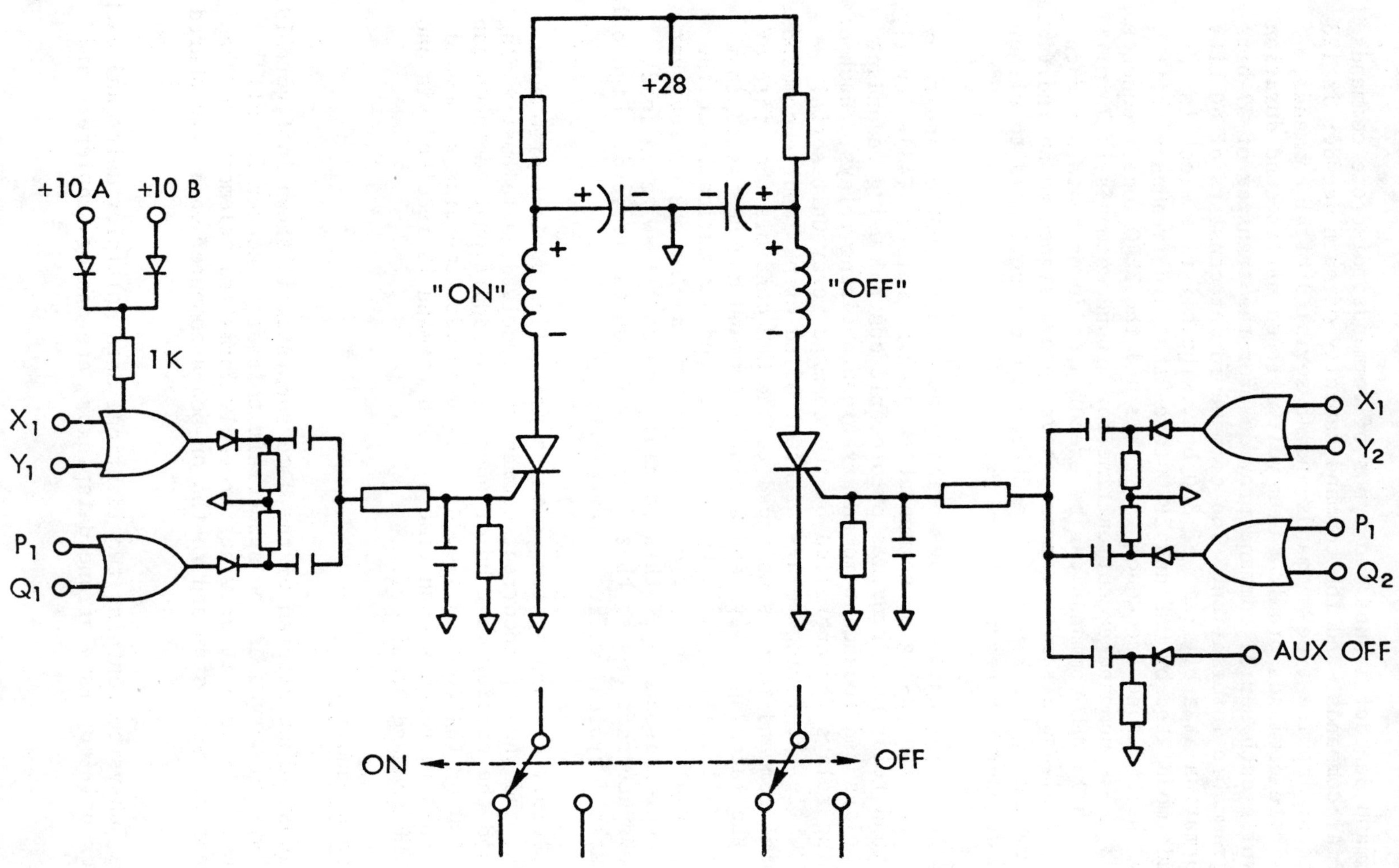

Figure 18. Relay Driver

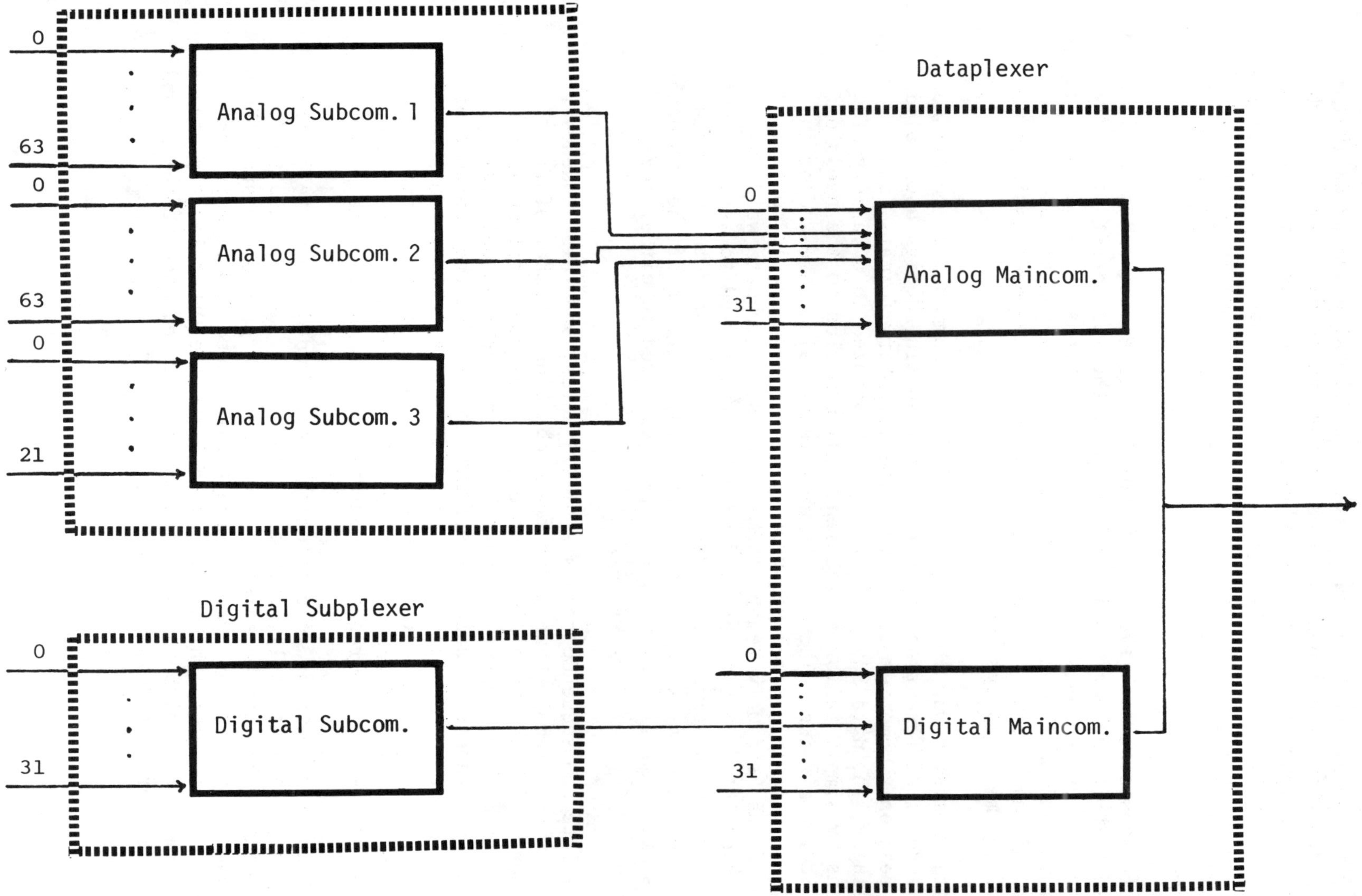

Figure 19. IUE Data Subsystem Block Diagram

one digital subplexer. The dataplexer has 32 analog and 32 digital input gates where the input to three of the analog gates are the subcommutated outputs of the analog subplexer and the input to one of the digital gates is the subcommutated output of the digital subplexer. The analog subplexer houses two 64-channel and one 32-channel subcommutators while the digital subplexer houses one 32-channel subcommutator.

The data subsystem was given access to all engineering and scientific data so that it can perform its function as the spacecraft telemetry encoder. It uses time shared techniques to alternately send data to the ground and to the on-board computer. Table 4 summarizes the number and the type of telemetry channels that have been assigned to each of the major subsystems.

The data subsystem was designed to satisfy the commonality requirements among several different spacecraft. The resulting design used solid state read only memories to generate the telemetry formats, and accordingly, the data subsystem was easily adapted to a variety of missions by simply tailoring the read only memories to the needs of a particular mission. The dataplexer selects main frame and/or subcommutated data in a timed sequence, and this sequence is controlled by one of the Read Only Memory (ROM) formats. Each data word is transformed into a 8 bit word and each is transferred to a serial data bit stream. Data words are alternately made available to the ground and to the OBC using time shared techniques, and each of the two data streams are controlled by ROMs that are dedicated to either the ground or to the OBC respectively.

One complete ROM sequence consists of 128 words, and this sequence is called a minor frame. Each minor frame contains a number of fixed parameter words and these always appear in the same location within the main frame independent of the contents of the read only memory. These fixed parameters include the frame sync words, the contents of the minor frame counter, and the spacecraft clock. Figure 20 shows a typical minor frame telemetry format. A major telemetry format is defined as 256 minor frames. Consequently, even the 64 word analog subcommutated words are repeated four times each major frame.

Each dataplexer contains four selectable hard coded ground telemetry ROM formats. Additionally, each contains one variable ground telemetry format memory that can be selected and coded via ground command. Each dataplexer also contains two selectable hard coded ROM formats and a variable format memory which are dedicated to the OBC telemetry stream. These formats control both the contents and the sequence of the ground and OBC telemetry streams.

<u>TABLE 4</u>

<u>IUE TELEMETRY CHANNEL ASSIGNMENTS</u>

	NUMBER OF INPUT LINES				
	DATAPLEXER			SUBPLEXER	
SUBSYSTEMS	ANALOG	DIGITAL	STATUS BITS	ANALOG	DIGITAL
Power	3		2	20	2
Command		2	8	6	
Communications			4	11	
Computer		4		5	6
Data Systems	5		3	2	
Scientific Instru.	6	6	2	25	10
Attitude Control	14	11	2	60	11
Hydrazine				29	1
Ordnance				2	2
Spacecraft			2		
Spare	1	8			
Digital Subcom.		1			
Analog Subcom.	3				
TOTAL	32	32		160	32

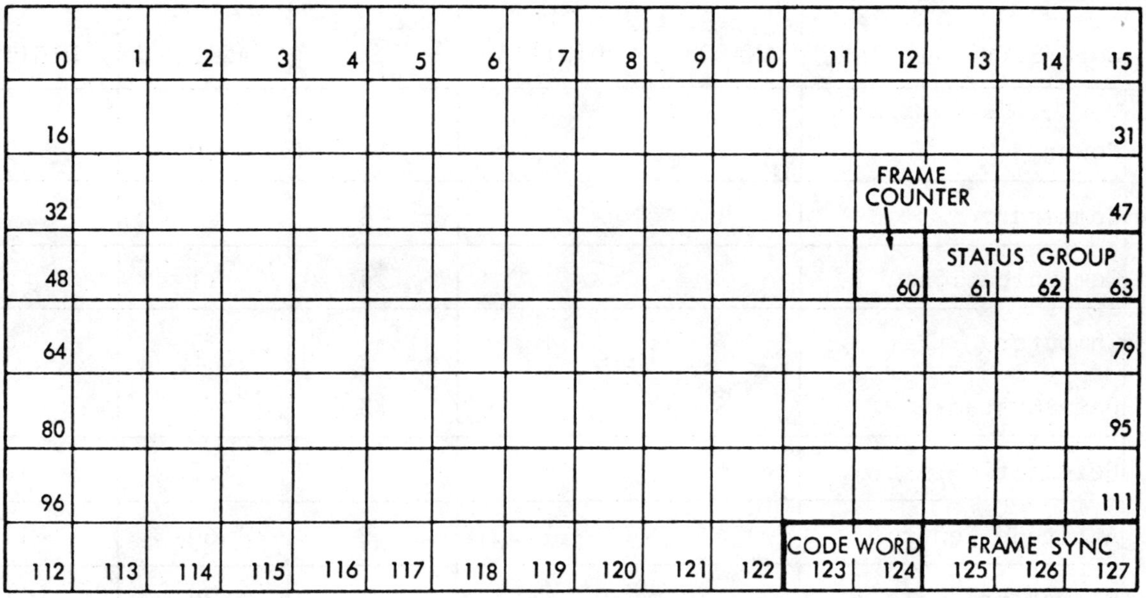

Figure 20. Dataplexer Telemetry Format

The telemetry bit rate is selectable via ground command from 1.25 kb/second to 40 kb/second in binary steps, and the multiplex ratio between OBC and ground telemetry can be varied from 1:1 to 32:1. This choice of multiplex ratios, for example, allows the selection of 40 kb/second for both ground and OBC data or the selection of 1.25 kb/second and 40 kb/second ground and OBC data respectively.

In an attempt to make the most effective use of the available data formats, each of the four selectable ROMs has been programmed to satisfy a specific need. The first afforded visibility of those parameters that were of particular importance to the ascent phase and early acquisition modes. This format also emphasizes engineering data by providing a rapid repetition rate for subcommutated data even at 1.25 kb/second, and provides a convenient format for use during the period of spacecraft checkout.

The second format is programmed as an operational format and this format is routinely selected during the mission orbit phase of operations except during camera image readouts. Format three is dedicated to quickly dumping a camera image and format four is a diagnostic format that is dedicated to quickly dumping the contents of the on-board computer plated wire memories. But even though formats three and four are special purpose formats, each is programmed to contain all of the subcommutated data so that even when the camera image or computer memories are being read out, all of the engineering data is provided on the ground as well. Figures 21 through 24 illustrate each of the ground data formats, and these figures illustrate an additional feature provided by the IUE data system--the ability to change even a hard wired ROM by using the two available indirect address registers per ROM to select a parameter for display. The indirect address register feature allows the selection of any one of sixteen digital or analog non-subcommutated parameters. Finally, the variable format memory can be used as a backup to any of the hard coded ROMs or it can be used for diognostic purposes. The on-board computer ROMs have been programmed to provide the OBC with those parameters that it requires to carry out its repertoire of attitude control and sundry other computations. As in the case of ground telemetry, a variable format memory is available for backup to the hard wired ROMs or for diagnostic purposes.

As was mentioned earlier, the data system was built with the goal of commonality in mind. Clearly its almost infinite versatility does not come without drawbacks. The redundant data multiplex units, the selectability of fixed and variable ROM formats, the choice of data rates and multiplex ratios, and the ability to use an indirect address capability created a virtual nightmare with regard to the data subsystem integration and test program. It was clearly impossible to test all of the possible combinations and we paid the penalty by learning, through

BIT RATE=1.25KB/SEC

AVERAGE SAMPLE TIME		
MF	0.819	SEC
ASC1	3.28	SEC
ASC2	3.28	SEC
ASC3	3.28	SEC
DSC	3.28	SEC
ESC	26.24	SEC

Figure 21. Transfer Orbit/Accelerated Engineering Data Format

0 S/C I	1 ACCEL A	2 ACCEL B	3 COMP ACCEL LEVEL	4 ANALOG SUBCOM 1 (64) #	5 #	6 #	7 #	8 #	9 #	10 #	11 #	12 ANALOG SUBCOM 3 (32) #	13 #	14 #	15 #
16 S/C I	17 BUSS V 27/29	18 SMSS #	19	20 ANALOG SUBCOM 1 (64) #	21 #	22 #	23 #	24 #	25 #	26 #	27 #	28 ANALOG SUBCOM 3 (32) #	29 #	30 #	31 #
32 S/C I	33 ACCEL A	34 ACCEL B	35 COMP ACCEL LEVEL	36 PITCH ERROR (CSS)	37 ROLL ERROR (CSS)	38 PAS/FES DATA **#	39 **	40 **	41 **	42 **	43 **	44 **	45 **	46 **	47 **
48 S/C I	49 BUSS V 27/29	50 DMU RAD MON	51 OBC DATA *#	52 *#	53 *#	54 *#	55 *#	56 *#	57 *#	58 CMDEX COUNT # 1	59 CMDEX COUNT # 2	60 FRAME COUNT	61 STATUS GROUP	62	63
64 S/C I	65 ACCEL A	66 ACCEL B	67 COMP ACCEL LEVEL	68 ANALOG SUBCOM 2 (64) #	69 #	70 #	71 #	72 #	73 #	74 #	75 #	76 EXPERIMENT SUBCOM 1(64) #	77 #	78 EXPERIMENT SUBCOM 2(64) #	79 #
80 S/C I	81 BUSS V 27/29	82 PITCH RATE 1	83 YAW RATE 1	84 ROLL RATE 1	85 PITCH RATE 2	86 YAW RATE 2	87 ROLL RATE 2	88 DIGITAL SUBCOM #	89 #	90 #	91 #	92 #	93 #	94 #	95 #
96 S/C I	97 ACCEL A	98 ACCEL B	99 COMP ACCEL LEVEL	100 PITCH WHEEL CMD	101 YAW WHEEL CMD	102 ROLL WHEEL CMD	103 S/C V 0-30	104 DIGITAL SUBCOM #	105 #	106 #	107 #	108 #	109 #	110 #	111 #
112 S/C I	113 BUSS V 27/29	114 DMU CAL	115 ANALOG SUBCOM 2(64) #	116 #	117 #	118 #	119 #	120 #	121 #	122 #	123 FRAME PARITY	124	125 FRAME SYNC	126	127

* INDIRECT ADDRESS REGISTER 1
** INDIRECT ADDRESS REGISTER 2
\# WORD GATE REQUIRED
MF MINOR FRAME

FRAME SYNC NORMAL HEX(FAF320)
FOR CONVOLVED HEX(SFBE)
CONVOLVED HAS ONLY 16 BITS SYNC CODE

BIT RATE=40KB/SEC

AVERAGE SAMPLE TIME		
MF	25.6	MSEC
ASC1	1.64	SEC
ASC2	1.64	SEC
ASC3	.82	SEC
DSC	1.64	SEC
ESC	1.64	SEC

Figure 22. Scientific Instrument Video Data Format

0	1	2	3	4	5	6	7	8	9	10	11	12	13	14	15
						CAMERA ANALOG VIDEO DATA									
**#	**#	**#	**#	**#	**#	**#	**#	**#	**#	**#	**#	**#	**#	**#	**#
16	17	18	19	20	21	22	23	24	25	26	27	28	29	30	31
						CAMERA ANALOG VIDEO DATA									
**#	**#	**#	**#	**#	**#	**#	**#	**#	**#	**#	**#	**#	**#	**#	**#
32	33	34	35	36	37	38	39	40	41	42	43	44	45	46	47
						CAMERA ANALOG VIDEO DATA									
**#	**#	**#	**#	**#	**#	**#	**#	**#	**#	**#	**#	**#	**#	**#	**#
48	49	50	51	52	53	54	55	56	57	58	59	60	61	62	63
ASC1	ASC2	ASC3				OBC DATA				CMDEX COUNT	CMDEX COUNT	FRAME COUNT		STATUS GROUP	
#	#	#	*#	*#	*#	*#	*#	*#	*#	# 1	# 2				
64	65	66	67	68	69	70	71	72	73	74	75	76	77	78	79
						CAMERA ANALOG VIDEO DATA									
**#	**#	**#	**#	**#	**#	**#	**#	**#	**#	**#	**#	**#	**#	**#	**#
80	81	82	83	84	85	86	87	88	89	90	91	92	93	94	95
						CAMERA ANALOG VIDEO DATA									
**#	**#	**#	**#	**#	**#	**#	**#	**#	**#	**#	**#	**#	**#	**#	**#
96	97	98	99	100	101	102	103	104	105	106	107	108	109	110	111
						CAMERA ANALOG VIDEO DATA									
**#	**#	**#	**#	**#	**#	**#	**#	**#	**#	**#	**#	**#	**#	**#	**#
112	113	114	115	116	117	118	119	120	121	122	123	124	125	126	127
DSC	S/C I	ESC1	ESC2				CAMERA STATUS				FRAME PARITY CODE WORD			FRAME SYNC	
#		#	#	**#	**	**	**	**	**	**					

* INDIRECT ADDRESS REGISTER 1
** INDIRECT ADDRESS REGISTER 2
\# WORD GATE REQUIRED
MF MINOR FRAME

BIT RATE=40KB/SEC

AVERAGE SAMPLE TIME		
MF	25.6	MSEC
ASC1	.41	SEC
ASC2	.41	SEC
ASC3	.82	SEC
DSC	.41	SEC
ESC	.82	SEC

Figure 23. Operational Data Format

0 *#	1 *	2 *	3 FES DATA *	4 *	5 *	6 *	7 DMU CAL RAMP	8 #	9 CAMERA1 STATUS	10 GYRO 1A #	11 GYRO 1B	12 GYRO 2A #	13 GYRO 2B	14 GYRO 3A #	15 GYRO 3B
16 *#	17 *	18 *	19 FES DATA *	20 *	21 *	22 *	23 DMU RAD MON	24 #	25 CAMERA2 STATUS	26 GYRO 4A #	27 GYRO 4B	28 GYRO 5A #	29 GYRO 5B	30 GYRO 6A #	31 GYRO 6B
32 *#	33 *	34 *	35 FES DATA *	36 *	37 *	38 *	39 PITCH ERROR CSS	40 #	41 CAMERA3 STATUS	42 BUSS I	43 BUSS V 27/29	44 #	45 DIGITAL SUBCOM #	46 #	47 #
48 *#	49 *	50 *	51 FES DATA *	52 *	53 *	54 *	55 ROLL ERROR CSS	56 #	57 CAMERA4 STATUS	58 CMDX COUNT 1 #	59 CMDX COUNT 2 #	60 FRAME COUNT	61	62 STATUS GROUP	63
64 *#	65 *	66 *	67 FES DATA *	68 *	69 *	70 *	71 EXPERIMENT SUBCOM-SYS1 #	72 # 64	73 #	74 #	75 #	76 OBC 1 DATA #	77 #	78 #	79 #
80 *#	81 *	82 *	83 FES DATA *	84 *	85 *	86 *	87 EXPERIMENT SUBCOM-SYS2 #	88 # 64	89 #	90 #	91 #	92 OBC 2 DATA #	93 #	94 #	95 #
96 *#	97 *	98 *	99 FES DATA *	100 *	101 *	102 *	103 #	104 ANALOG SUBCOM 1 #	105 # 64	106 #	107 #	108 ANALOG SUBCOM 2 #	109 # 64	110 #	111 ANALOG SUBCOM3 # 32
112 **#	113 **	114 **	115 **	116 FES/PAS DATA **	117 **	118 **	119 **	120 **	121 **	122 BUSS V 0,30	123 FRAME PARITY CODE	124	125	126 FRAME SYNC	127

* INDIRECT ADDRESS REGISTER 1
** INDIRECT ADDRESS REGISTER 2
\# WORD GATE REQUIRED
MF MINOR FRAME

BIT RATE=40K. SEC

AVERAGE SAMPLE TIME		
MF	25.6	MSEC
ASC1	1.64	SEC
ASC2	1.64	SEC
ASC3	.82	SEC
DSC	.82	SEC
ESC	1.64	SEC

Figure 24. On-Board Computer Memory Dump Format

0	1	2	3	4	5	6	7	8	9	10	11	12	13	14	15
							OBC MEMORY DUMP (DIGITAL)								
**#	**	**	**	**#	**	**	**	**#	**	**	**	**#	**	**	**
16	17	18	19	20	21	22	23	24	25	26	27	28	29	30	31
							OBC MEMORY DUMP (DIGITAL)								
**#	**	**	**	**#	**	**	**	**#	**	**	**	**#	**	**	**
32	33	34	35	36	37	38	39	40	41	42	43	44	45	46	47
							OBC MEMORY DUMP (DIGITAL)								
**#	**	**	**	**#	**	**	**	**#	**	**	**	**#	**	**	**
48 ASC1 #	49 ASC2 #	50 ASC3 #	51 *#	52 *#	53 *#	54 OBCDATA *#	55 *#	56 *#	57 *#	58 CMDEX COUNT # 1	59 CMDEX COUNT # 2	60 FRAME COUNT	61	62 STATUS GROUP STATUS GROUP	63
64	65	66	67	68	69	70	71	72	73	74	75	76	77	78	79
							OBC MEMORY DUMP (DIGITAL)								
**#	**	**	**	**#	**	**	**	**#	**	**	**	**#	**	**	**
80	81	82	83	84	85	86	87	88	89	90	91	92	93	94	95
							OBC MEMORY DUMP (DIGITAL)								
**#	**	**	**	**#	**	**	**	**#	**	**	**	**#	**	**	**
96	97	98	99	100	101	102	103	104	105	106	107	108	109	110	111
							OBC MEMORY DUMP (DIGITAL)								
**#	**	**	**	**#	**	**	**	**#	**	**	**	**#	**	**	**
112 DSC #	113 DSC #	114 ESC1 #	115 ESC2 #	116 DMU CAL RAMP	117 DMU RAD MOM	118 BUSS VOLTS 0-30V	119 DMU 2-5V CAL2M	120 27-29 VOLTS BUSS	121 DMU 2-5V CAL	122 S/C BUSS I	123 FRAME PARTY WORD	124 FRAME PARTY WORD	125 FRAME SYNC	126 FRAME SYNC	127 FRAME SYNC

* INDIRECT ADDRESS REGISTER 1
** INDIRECT ADDRESS REGISTER 2
\# WORD GATE REQUIRED
MF MINOR FRAME

in-orbit operations, that some of the data rate and multiplex ratio combinations caused problems with regard to OBC operations. These problems were resolved by imposing ground operational guidelines and constraints.

4.3.3 Computer

The IUE on-board computer is the Goddard developed Advanced On-board Processor (AOP). Its predecessor was successfully flown on the Orbiting Astronomical Observatory and on other Goddard missions. Once again, commonality was a key design concern, and the computer was designed to accept a Special Input Output (SIO) unit that would adapt the AOP to accommodate the needs of various missions.

The on-board computer consists of redundant processor modules and three plated wire memory modules each of which is 4096 words in size. The processor module contains a dedicated power converter, the Central Processing Unit (CPU) and the special interface circuitry used for input/output. Although the special interface circuitry is mission unique, it is packaged within the processor module so that each processor is a stand alone unit that only requires memory.

Data words and instructions are 18 bits in length and negative numbers are represented in two's complement form. The instruction repertoire contains 55 instructions and these include a 5 microsecond add, a 37 microsecond multiply, and a 75 microsecond divide.

The basic memory module for the OBC is a random access 4096 by 18-bit word, plated wire unit, and three of these modules have been utilized for the IUE application. The memories feature cycle by cycle power switching, have a 500 nanosecond access time, and a 1 microsecond cycle time. They are nonvolatile and the readout is a nondestructive process. The on-board computer is the heart of the IUE system design. It performs all attitude control algorithm computations and issues all reaction wheel torquing commands. It performs self checks, monitors spacecraft health and safety functions, accurately controls camera exposure times, and provides a delayed command capability. The complete flight software program resides in two of the memory units while the third contains the executive and the basic control system hold and slew algorithms only. As indicated earlier, the computer receives its data via the dataplexer, and in addition, the OBC inserts data into the telemetry stream just as any other data source would do. Either dataplexer can communicate with either of the redundant OBC processors, and either processor can communicate with either command decoder.

4.4 STABILIZATION AND CONTROL

The stabilization and control subsystem is without question the most complex subsystem aboard the IUE spacecraft. It was designed to be simple, light weight, low powered and reliable on one hand, and on the other hand, it was required that the control subsystem satisfy precise pointing and slew accuracy specifications. Many alternate designs were considered before a single viable approach was pursued in detail. Deliberate attempts were made to use proven design concepts, to utilize flight qualified or readily available hardware, to draw from the experience of previous and ongoing Goddard programs of a comparable nature, and to achieve component commonality with other Goddard programs wherever practical.

In addition to the general requirements that have already been stated, the principal requirements that have largely influenced the control subsystem design philosophy are the following:

a. Three-axis stabilization in inertial space with a ± 1 arc second pointing accuracy requirement in pitch and yaw, for periods typically of 1/2 hour and more.

b. Following a slew or a sequence of slew maneuvers, a new source target must be acquired to within the 8 arc-minute half-cone angle field of the telescope.

c. The pointing accuracy requirement must be satisfied for extended periods of time even when stellar attitude measurements of sufficient accuracy and frequency are not available.

d. The expected useful operating life was to be 3 to 5 years.

e. The stabilization and control subsystem was to be autonomous from a safety standpoint even though the spacecraft would be in synchronous orbit and in nearly continuous contact with the ground.

When attempting to satisfy the stated requirements, it was necessary to consider design solutions with regard to four basic areas of concern; these include torquing devices, attitude sensors, the logic circuitry and electronics needed to mechanize the control laws, and the method of information flow that was to be used between sensors and torques.

The control concept that was considered most suitable for the IUE was a momentum exchange control system designed on a single-axis basis using reaction wheel control. Three axis stabilization in inertial space implies that the three control axes are essentially decoupled. Clearly some gyroscopic coupling develops when the spacecraft is slewed even when slewing is performed on a single axis basis; however, the holding accuracy requirement in the other two axes can easily be relaxed during such slew maneuvers.

The sensor complement selected includes Earth and Sun sensors for use in ground computer aided attitude determination, and accelerometers for on-board nutation control as well as rate gyro sets, and analog Sun sensors for initial spacecraft acquisition. An Inertial Reference Assembly, fine Sun sensors, and fine error sensors (or star trackers) are used for inertial star acquisition and subsequent hold and slew operations. Hardwired analog control algorithms were incorporated to perform nutation control and all spacecraft initial acquisition functions as well as to provide an emergency rate hold function. The on-board computer was programmed to perform all the hardwired algorithms in addition to its primary attitude control functions of star acquisition, hold, and slew. Attitude control authority is provided by the hydrazine propulsion system or by any three of the four momentum reaction wheels.

The choice of attitude sensors was somewhat involved. Other things being equal, a cooperative target is always the easiest to track for any pointing control system primarily because of the self-calibrating nature of the system. The optical design of the IUE, however, precluded the possibility of using the target star as the attitude control system target. This is true because the optical design requires that the target drop through an aperture hole before it can be imaged onto the optics of the scientific instrument; once this occurs, the target is no longer available to the control system for error signal generation. Accordingly, an offset pointing method of control was selected where one or more guide stars from the star field surrounding the target star are used to generate the control system error information.

The IUE control laws incorporate two alternate methods of offset guide star control. The first requires a bright guide star, and in this mode the fine error sensor or star tracker is operated in its track mode to produce pitch and yaw axis error signals directly. The second method of control allows the use of faint guide stars. The second mode of operation requires that rate integrating gyros supply three axis attitude while the tracker is used to periodically reference the pitch and yaw axis gyros. The gyros are used to impart three axis rate damping information for both modes of operation. Sources that are eleventh magnitude and brighter are used directly as error sources while sources fainter than eleventh magnitude are used in the gyro update mode.

Analyses demonstrated the requirement of a very precise, well calibrated, gyroscopic inertial measurement system that is capable of maintaining an attitude reference to sub-arc second accuracy for several minutes of time in pitch and yaw. Independent of this analysis, a study of the slew mode requirements led to a similar conclusion about the need of a precise inertial reference system, albeit with somewhat relaxed accuracy specifications. In the absense of a gyroscopic system, well separated guide stars must be sensed by two star trackers to maintain inertial attitude information throughout slew maneuvers.

Single-degree-of-freedom, hydrodynamic gas bearing, rate-integrating gyroscopes with pulse-rebalance electronics were chosen to meet the IUE requirements. The gas bearing effectively eliminates the gyro bearing failure mechanism and significantly reduces its output noise, while pulse rebalancing provides a much improved readout accuracy as compared to analog rebalancing techniques. These gyros are used in a strapdown mode. While spacecraft such as the OAO have demonstrated that good, high quality gyros can operate for at least 60,000 hours (to date) in orbit without failures or degradation, prudence dictated the use of redundancy in hardware to ensure that the IUE 3 to 5 year operating goal is met. The usual method of employing redundancy would have involved the use of two, three-axis gyro packages with their input axes aligned to the spacecraft control axes. This type of redundancy is often improved by providing a cross strap so that any combination of gyros in either package can be used to permit control system operations even in the event of gyro failures in both systems. However, even with this type redundancy, control about a particular axis is lost if both gyros which sense that body axis fail.

The mechanization of gyro redundancy on IUE involves the use of six individual gyroscopes mounted in a single package. The input axes of each of the gyros are skewed relative to each other and to the spacecraft control axes. As a result, each gyro senses some component of body rate about each of the spacecraft control axes. Accordingly, if the outputs of any three gyros are properly combined, three axis information can be reconstructed. With this scheme of redundancy, as many as three gyros can fail without seriously impacting system performance and accuracy.

A hydrazine system was primarily required to adjust the initial orbit as well as to maintain orbit control throughout the life of the satellite. This system also provided the torquing authority needed for the acquisition modes and for maintaining spacecraft attitude during orbit control burns in the presence of orbit-control-thruster misalignments. Additionally, it was used to counteract the secular reaction wheel momentum build up that is caused by the integrated effects of the environmental disturbances. Computations showed that neither magnetic nor gravity gradient unloading techniques are useful for synchronous orbit missions.

The question of reaction wheel reliability was raised because of the 3 to 5 year lifetime requirement. Although wheels have demonstrated a high degree of reliability on many satellite programs, it was decided that a skewed redundant wheel would be flown to provide some added insurance. This wheel is skewed symmetrically with respect to each of the three orthogonal control axes.

The stabilization and control subsystem is fully redundant and was used to help attain the correct orbit as well as to control the pointing and slewing of the telescope. Just prior to third stage separation, the spacecraft was spin stabilized in the transfer orbit about its minimum moment of inertial axis. Because of this, it tended to nutate in the presence of on-board energy dissipation sources. Accordingly, the control system used a hydrazine engine pulsed through an on-board algorithm as an active nutation damper to prevent a large nutation angle build-up. The algorithm was driven by a linear accelerometer and it maintained the nutation half-cone angle below one degree. During the transfer orbit the control subsystem was also necessary to precess the spacecraft spin axis 180° in preparation for the apogee engine burn. Spacecraft attitude was calculated on the ground using the Sun and Earth sensor telemetered data. Once the spacecraft achieved geosynchronous orbit, a despin sequence was completed using coupled pairs of hydrazine engines. The solar arrays were deployed and the spacecraft was commanded to automatically acquire the Sun. This Sun acquisition sequence was accomplished using analog Sun attitude sensors and rate information derived from the on-board gyro system. Alignment requirements for this acquisition mode are controlled to better than one degree with respect to the spacecraft Sunline.

The stabilization and control subsystem is required to perform a wide variety of functions; however, the scope of this presentation allows for a detailed discussion of only the primary hold-slew operational mode.

Figure 25 defines the spacecraft control system axes and figure 26 presents a functional system block diagram. These figures depict the sensor systems upon which the control system depends to perform its precision hold and slew maneuvers. The sensor systems include: (1) a set of six gas-bearing, pulse rebalanced, inertial grade gyros each providing 0.01 second-of-arc integrated rate resolution for rates up to 600 seconds-of-arc per second, (2) a redundant two-axis digital Sun sensor system which provides angular resolution to 15 seconds-of-arc over a field-of-view of 64 X 124°, and (3) a redundant set of star trackers mounted within the scientific instrument and utilizing the telescope optics. Each tracker is capable of creating a star field map or of performing a search and track operation. While in the star mapping mode, the tracker telemeters the intensity and the location of all sources within its 16 arc-minute field of view. Alternately, the tracker can be operated in a tracking mode where it is commanded

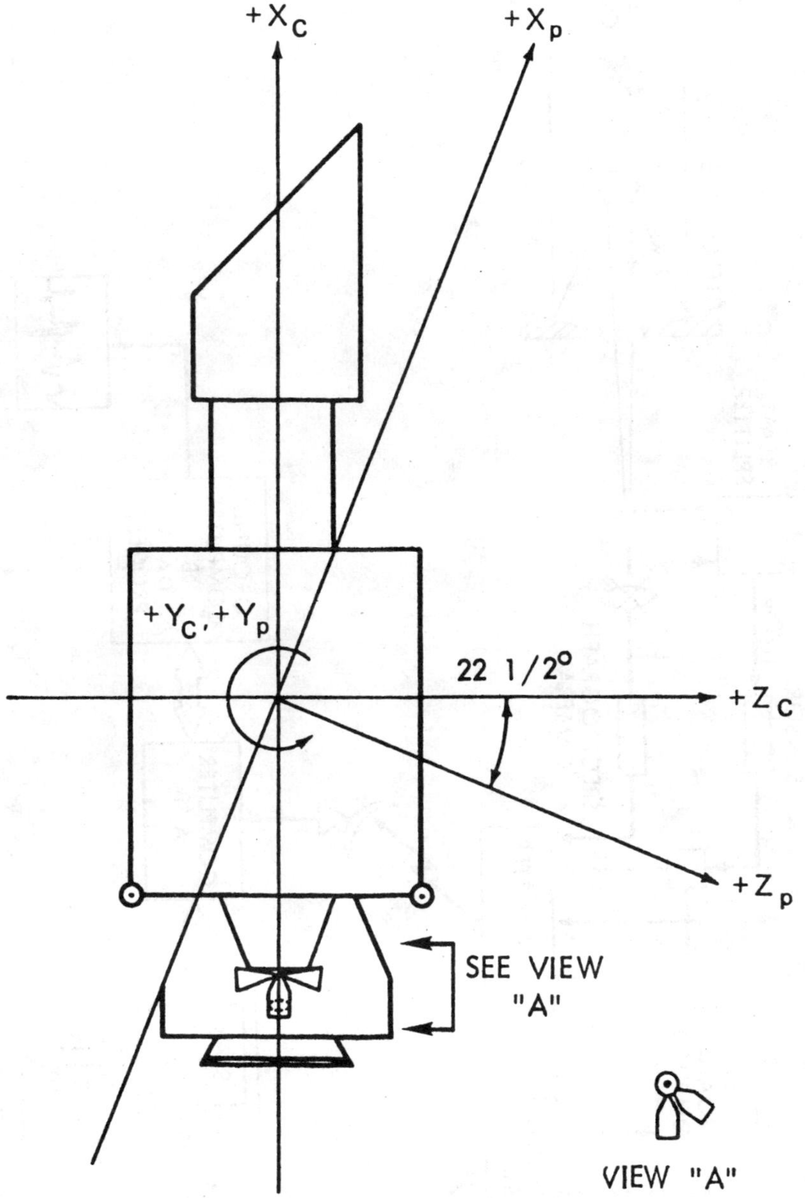

Figure 25. Axis Definition for IUE Controls System

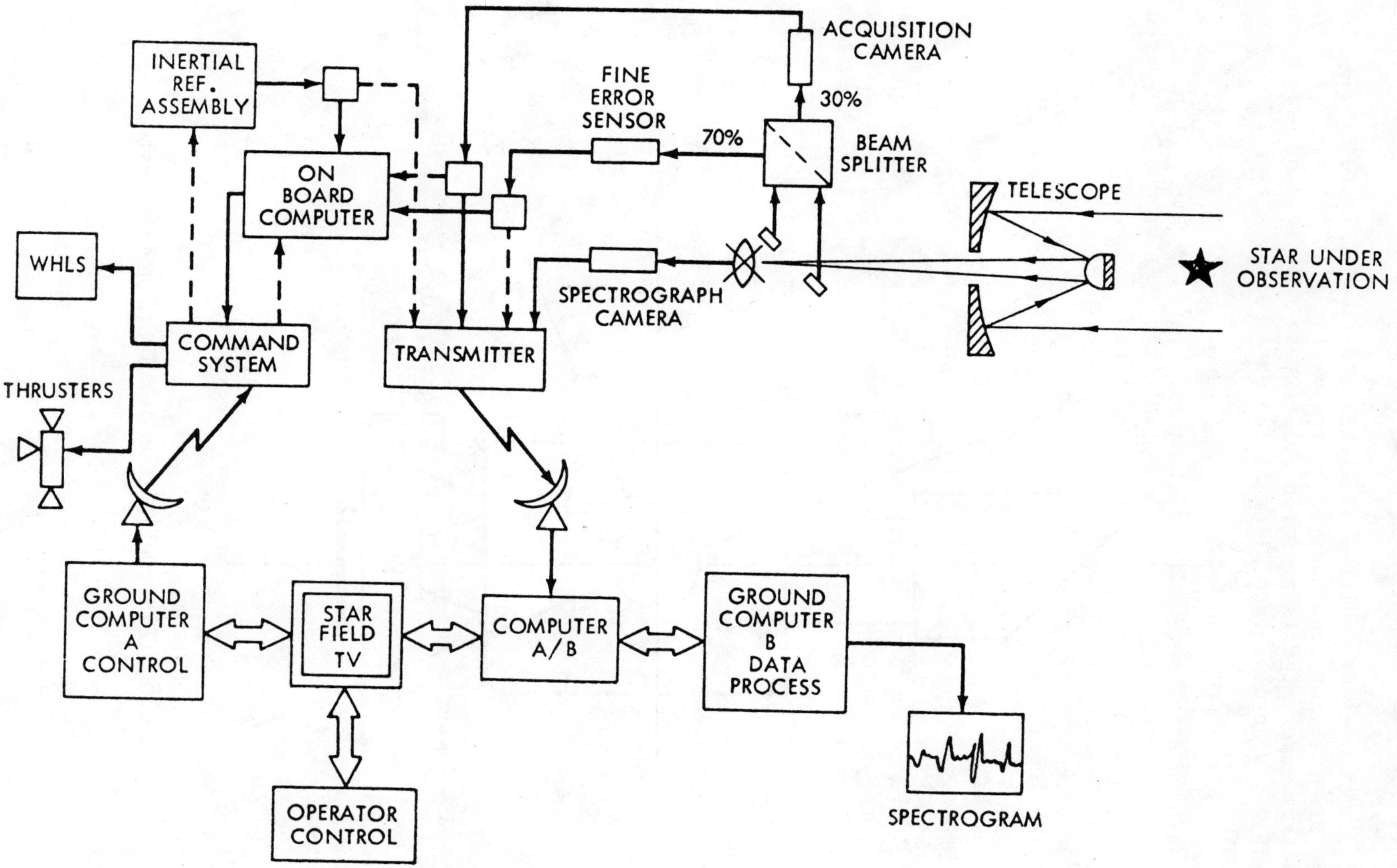

Figure 26 . Block Diagram of IUE Essential System Elements

to lock onto a particular star within its field. In this mode, the fine error sensor continues to telemeter the star intensity and the two axis error signals measured relative to the field of view center point. The angular resolution of the error signals is 0.27 seconds of arc.

A guest observer requires that the spacecraft telescope axis be slewed from one target source to another, and although the on-board systems do not know the absolute inertial reference of the spacecraft, the control center ground computer is inertially re-referenced each time a new target is acquired. This referencing is possible because the precise location of all target and some guide stars can be found using catalogues of ground-based-telescope generated star maps. The system is referenced initially by creating a number of star maps and by using star pattern recognition techniques.

To slew from one target to another, the observer is only required to enter the location of his new target into the control center computer. The ground software takes this information and computes a slew maneuver in gyro coordinates. This maneuver sequence is displayed and the resident astronomer initiates the uplink of the computed gyro biases to the on-board computer (see figure 27). The maneuver is executed upon receipt, and once the transient has settled, the operator commands the fine error sensor to generate a star map of its new field of view (see figure 28). The guest astronomer studies this map in real time and selects his target from other stars that might also be presented within the star field using a display driven curser (see figure 29). The computer generates a second slew maneuver and this maneuver places the target source into the aperture of the scientific instrument. The fine error sensor will then be commanded to track on a guide star, a star in the vicinity of the target star, and the FES attitude error information will allow the control system to maintain the target within the aperture hole.

For each of the maneuvers described the slew is controlled by the three axis gyro set. The fine error sensor provides an accurate gyro position and drift trim reference for the spacecraft pitch and yaw axes while the fine Sun sensor is used to provide these references for the spacecraft roll axis.

Once the target star is within the 3 arc-second experiment aperture, the control system must maintain this inertial position for periods of typically 1 hour or more. If the tracker is locked onto a bright guide star during this period, the pitch and yaw axis position control is accomplished using tracker position errors directly. When only a faint guide star is available to the tracker, control is maintained using the gyro generated position information, but here the FES error

Figure 27. Resident Astronomer Uplinks Gyro Biases

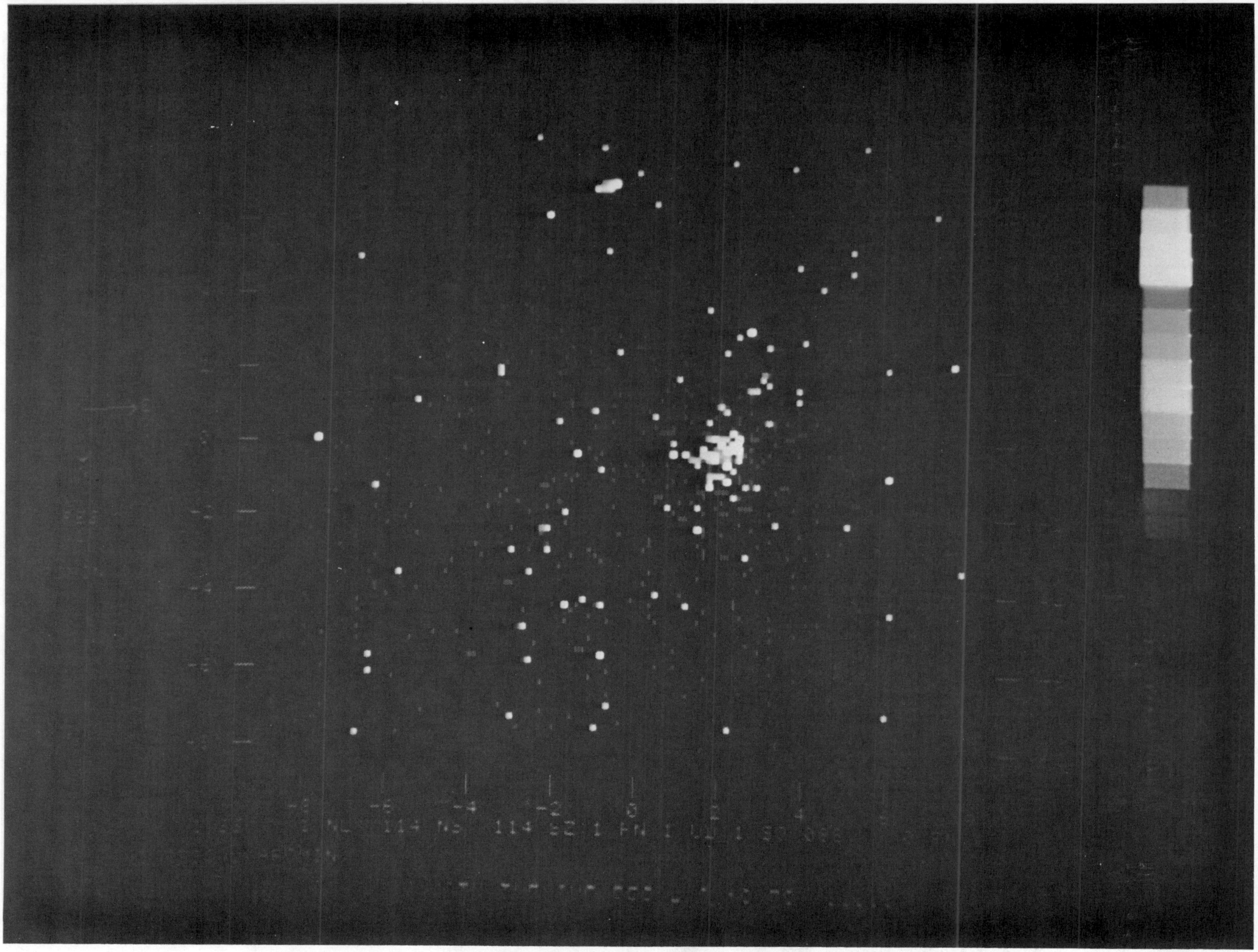

Figure 28. Fine Error Sensor Generated Star Map

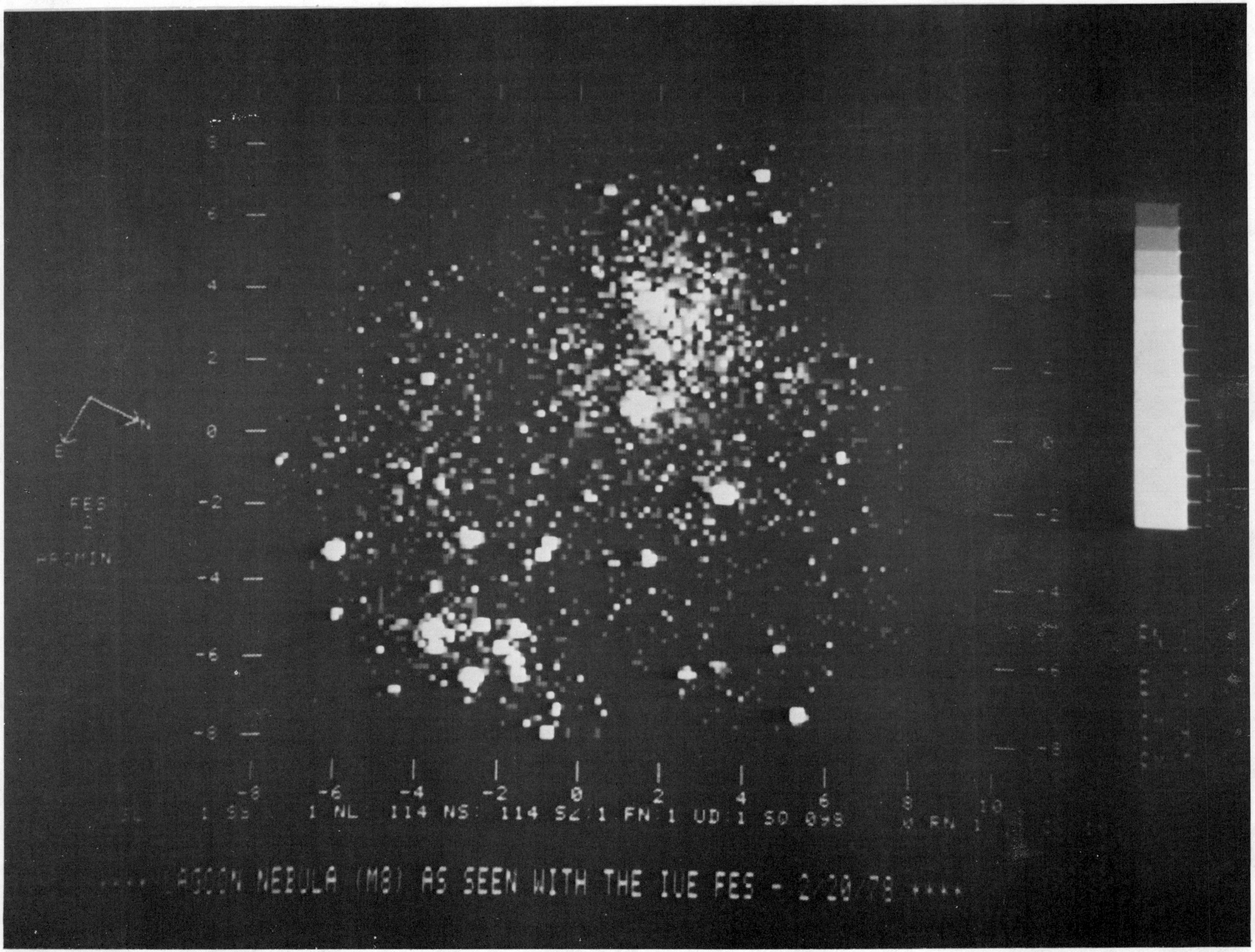

Figure 29. Selection of Target Star

signals are used to effect periodic gyro position and rate trims. Finally, if no guide star appears within the telescope field of view, the gyros must be well trimmed using the closest available source, and then control must be maintained using only the gyro generated error signals.

The control laws are implemented digitally within the on-board computer using Kalman filtering techniques, and the on-board computer issues reaction wheel commands to the wheel drive circuitry five times a second. The wheels are driven in pulse modulator fashion where the torquing commands produce wheel torques that are proportional to the attitude error signals.

An attitude control system as complex as that of IUE does not conveniently lend itself to analysis; consequently, after a nominal parameter set was selected for each mode of operation, a digital computer simulation was used to evaluate system performance and to iterate the nominal parameter set as required. Through simulation methods, it was possible to accurately model the flight hardware, including the system nonlinearities and the spacecraft nonrigid body elements as required. Furthermore, it was possible to functionally represent the on-board computer software as it is coded within the on-board computer. Proper representation of the various samplers and sample-and-holds associated with the OBC iterative loops and input registers was also possible. Finally, it was possible to simulate the operation of the attitude control system in a changing environment. This feature allowed the verification of dynamic performance as the ground commands changed the mode, controller, sensor, and torquer configuration during a given computer simulation run.

To gain confidence in the integrity of this complex simulation, a two axis "fine pointing" test was performed that used a fully integrated set of attitude control hardware, a driven servo table, and a remote sensor target. The two reaction wheels were specially modified by attaching a shaft position resolver to each of their rotors, and the position error signals were then used to accurately drive the table upon which the spacecraft hardware was mounted. Wheel rotations were converted to table rotations through the use of gear trains where the gear ratios were tailored to represent the ratios between the pitch and yaw body inertias and the pitch and yaw axis wheel rotor inertias. Test and simulation results compared favorably.

5. MAJOR PROBLEMS AND THEIR SOLUTIONS

Having described the design of the overall system and the major subsystems on IUE, we would like to relate some of the major problems encountered during the program. Some of the problems were unique, some were classic; however, the methods of analyzing and solving the problems are typical for research and development programs.

The initial problem we faced, and one that persisted for nearly the entire prelaunch phase, was allowable spacecraft weight. As previously mentioned, the synchronous orbit was chosen for operational conveniences; unobstructed observing time was maximized and the operational concept would be familiar to the astronomers. However, the nominal synchronous transfer orbit capability of the Delta 2914 launch vehicle limited the spacecraft weight to 700 Kg. At the initiation of the flight project phase in July 1973, the estimated spacecraft weight was 667 Kg, resulting in only a five percent contingency. We knew this to be an extremely small margin and that considerable efforts would be required to maintain a viable program.

The weight problem was compounded by two other factors. IUE was to be a single mission program; no back-up launch was planned in the event of an early mission failure. Also, the mission lifetime objective was 3 years. These factors dictated a systems design with subsystem redundancy. In essence, we were planning to fly two of everything.

The structural design of the spacecraft stressed light weight materials. The lower cone structure which interfaced directly with the Delta launch vehicle was machined from a berylium alloy to maximize the strength/weight ratio. Magnesium was used liberally for electronics packaging frames. Graphite impregnated fiberglass was used for the longitudinal struts in the spectrograph section of the scientific instrument both because of its low weight and its low thermal expansion coefficient. Stringent weight, volume, and power budgets were allocated to each of the various subsystems.

As the program progressed from the design stage to hardware fabrication, we periodically revised the weight estimate. Invariably, the weight estimate would increase and result in a review of each subsystem to identify possible weight reduction schemes. A computer program was written that included every item in the spacecraft at the black box level. The program would compute the apogee motor fuel needed to accommodate a hardware weight change. Each item was identified as either an actual, measured weight, or an estimate. Each estimated weight was given a confidence factor depending on the development status of that particular component. This allowed us to spot the areas that warranted the most attention and offered the greatest potential for weight reductions. An item identified for possible weight reduction had to be analyzed for impact to program cost and schedule.

More than 50 items were investigated for possible weight savings during the course of the program, and 21 were adopted. The following are a few of the more significant items.

As previously mentioned, one of the program requirements was that the spacecraft be in continuous view of the U.S. ground station and in view of the European station at least 10 hours each day. The original orbit design was a conventional, circular synchronous orbit that required an apogee boost motor (ABM) with 272 Kg propellant and a total weight of 296 Kg. By redesigning the orbit to satisfy the minimum station coverage conditions, a lower energy, eccentric synchronous orbit was achieved with an ABM containing 218 Kg of propellant and a total weight of 237 Kg. Obviously the total difference in weight of the two ABM's was not available as useful reserve in that a larger spacecraft weight required more propellant to achieve the mission orbit. For the IUE orbit, approximately two-thirds of the total ABM weight differential was available to the fueled basic spacecraft. This was the largest weight saving scheme that we adopted.

Another major configuration change to reduce weight involved the addition of a second Fine Error Sensor (FES) in place of two Acquisition Camera Modules (ACM). The preliminary spacecraft system design included a single FES to be used as an absolute position reference by the stabilization and control subsystem and two ACM's to perform the star mapping function required for target star acquisition. The ACM's were to be delivered as part of the UK supplied hardware, but the UK had run into hardware and schedule problems and the estimated power and weight of the acquisition cameras had increased considerably. The FES development was progressing well during this same time period, and in fact, its performance characteristics had been improved sufficiently so that the FES could be used for the rapid creation of star field maps in addition to its originally intended star tracking capability. Accordingly, the rationale for the choice of sensor configuration was reconsidered. The entire control system operation and the target star acquisition process were restudied, and all interfaces such as mechanical, optical, electrical, and thermal as well as schedule and cost aspects were evaluated.

The total sensor requirements associated with both the FES and the ACM included reliability, the ability to rapidly create star field maps, and the capability to provide absolute inertially referenced error signals in pitch and yaw spacecraft coordinates. It was demonstrated that the FES could both create star field maps and supply error signals directly while the ACM could only conveniently satisfy the star mapping requirement. It could be used to produce error signals only through the use of a complex ground computer software package. As a result of our findings and because of our severe weight problem, we chose to fly two FES's and to delete both ACM's. This change resulted in an improved reliability and a direct weight savings of 7.6 Kg.

A substantial weight savings was also realized in the power system. The original design specified two 12 ampere-hour batteries to handle the spacecraft loads during the transfer orbit, the eclipse periods of the mission orbit, or when spacecraft loads exceed the solar array output. In order to reduce spacecraft weight, two changes to the battery configuration were investigated. The first was to consider a single 12 ampere-hour battery.

During the systems definition phase of the project, it was considered desirable to plan for normal spacecraft operations during the eclipse periods. This would allow spectrograph exposures in the absence of stray light and exposures closer to the Sun than 45°. But as the program progressed, it was realized that it would be impractical or at least inconvenient to constrain the orbit to accommodate observations in shadow due to the relatively short shadow time involved. Furthermore, analyses and tests indicated that stray light would not present any problem for sunlit observations. Consequently, it became reasonable to consider flying a reduced capacity battery even though a single battery configuration would only support essential spacecraft loads during eclipse periods. Accordingly, the orbit constraints were modified to yield the fewest and shortest duration eclipse periods possible. As the analysis of a single 12 ampere-hour battery configuration progressed, it became apparent that a better solution would be two batteries of 6 ampere-hours each. This configuration would impose a minimum of change requirements on the complete power system design and also eliminate the single point failure possibilities of a single battery configuration. Minimizing changes to the rest of the power system would also minimize schedule impact. The single battery configuration would have required ACS preparation, such as biasing the yaw wheel, prior to each eclipse in case of a battery failure during the shadow period. Critical firing of pyros during the transfer orbit was also a concern even though the odds were remote that the single battery would fail open. Although the single battery configuration would have resulted in a larger weight reduction, the two battery configuration was chosen because it offered more reliability, operational simplicity, and spacecraft safety.

Limited weight and power created a problem with the Inertial Reference Assembly (IRA) electronics. In order to meet specification, the successful bidder for the IRA contract proposed to use company developed hybrids in the electronics. While the company had demonstrated the capability of producing hybrids on other programs, some of the units required for the IRA involved a higher density of chips than they had previously produced. The new units were thought to be a "simple extension" of previous technology. This proved to be anything but the case for the IRA hybrids. At one point we investigated a second source for the entire IRA. We learned of another contractor who had produced a similar, but less sensitive, unit for another program. Spare hardware was available from the program that could be used for IUE with relatively minor modifications. We performed an interface study and had decided to commit contingency funds to this effort. However, the cameras continued to be the pacing item and we determined that the original IRA contractor would solve the hybrid problems without the need to contract for a second IRA.

Another major problem area involved the late delivery of certain subsystem flight hardware and the resulting impact on the total flight schedule. The development of the spectrograph cameras is a classic example of the types of problems one encounters in research and development programs.

During the systems definition phase, prototype vidicon tubes and UV converters for the spectrograph cameras were purchased for evaluation. Satisfactory performance was achieved in each case prior to initiation of the flight hardware phase. The vidicon tubes were ruggedized versions of a tube developed for ground use. The UV converter was a modified image intensifier that was developed for low light level applications.

As is frequently the case, the modification of the UV converter required more art than science, and the original supplier of the prototype devices was unable to manufacture flight quality units. The UK worked diligently with the US firm in an attempt to solve the many problems, particularly since a second source did not exist. The impact on the camera delivery dates became the largest single schedule problem in the entire program. Using some of the contingency funds, the GSFC project office initiated action to develop a second source. The UK-developed specifications were used to contract with a second US firm for a "best efforts" trial production. The initial results were extremely encouraging so an additional order was placed for a small number of flight quality converters. Again the second source responded and demonstrated that they could produce converters that met or exceeded all specifications. At this point, the UK dropped relations with the original supplier and contracted with the newly developed second source for flight converters.

The combination of the converter problems, potting problems and high voltage power supply problems resulted in the cameras being delivered to the project 24 months behind schedule. We managed to be ready for launch a year behind schedule, which means we absorbed half of the camera schedule delay, even though we eliminated most of the camera level calibrations in favor of scientific instrument level calibrations.

We employed many management techniques while attempting to solve the camera system problems. Monthly status reviews were held alternately in the UK and at the GSFC. We sent numerous GSFC experts in high voltage power supplies, materials, analog circuit design, and systems testing to the UK to serve as consultants. We used trend charts in an attempt to bound the delivery problem. The trend chart is a plot, by month, of the expected delivery date against the current date. Connecting the points representing expected delivery dates can be an indication of the ultimate delivery date. We also began systems functional and environmental testing using the flight spacecraft and the engineering unit scientific instrument as close to the original schedule as possible in order to minimize the test time that would be required after delivery of flight cameras. The engineering unit scientific instrument contained a single prototype camera, and it enabled us to solve several interface problems while awaiting flight cameras.

6. WHAT DID WE LEARN?

As mentioned earlier, IUE was handled as an in-house project with the GSFC as the "prime contractor." We let well over 100 contracts with private industry ranging from a few thousand to one and a half million dollars in value for certain subsystem hardware and software support items. All but one of our subsystem hardware contracts was let on a fixed price basis. We had a 20 percent funding contingency at the beginning of the program and 10 percent of the contingency funds remained at launch time to be available for the postlaunch phase. Less than five percent of our contingency funds was spent on increased hardware costs. This means that we had very good specifications and well qualified technical officers monitoring the contracts.

With the year's delay in the launch date, we did overrun the in-house manpower budget. As previously mentioned, the flight unit spectrograph cameras were delivered 2 years late. The original plan specified delivery of fully calibrated flight cameras 18 months before launch. We were anticipating a lengthy integration and test period to familiarize the operations teams with the complete observatory. Qualification of the short wavelength portion of the spectrograph required operation of the scientific instrument on an optical bench under vacuum conditions. Over a quarter of a million lines of code were written for the software to operate the observatory, and a lengthy software checkout period was anticipated. On several occasions, following significant slips in the camera delivery schedule, we considered shutting down the program at the GSFC, redeploying the personnel, and then resuming activities after flight camera delivery. However, as mentioned earlier, we opted to begin systems integration and testing using the flight spacecraft and the engineering unit scientific instrument in order to shorten the final testing phase with flight cameras. In this manner, we were able to launch just 7 months after delivery of flight cameras even though we had substantially shortened calibration of the cameras at the subsystem level in favor of extensive calibration at the full spacecraft systems level. We are convinced that creating a hiatus in the program would have been very inefficient and would have resulted in a longer launch date slip. A major factor in our success was keeping the in-house team together through launch and the in-orbit commissioning of the spacecraft.

On three occasions, we were impacted by the failure of single source suppliers to deliver components on schedule. Obviously one should avoid this situation on flight programs; however, it is difficult to do when you are pushing the state-of-the-art to fly the best possible research mission. It is imperative that the project office closely monitor the single source suppliers and be prepared to commit contingency funds to develop second sources when necessary.

A large fraction of our energies and resources was spent trying to control the overall spacecraft weight. During one of the campaigns, we persuaded the Engineering Directorate to designate a cash award for the best weight reduction suggestion that was adopted. At the time, the award was given for the suggestion to halve the battery capacity. For research programs of this nature, a weight margin of 15 to 20 percent should be established at the start.

In the final analysis, our success can be attributed to having many exceptionally competent and dedicated subsystem and systems-level personnel on the program. We maintained great team spirit, in the face of many crises, over a full 4 1/2 year prelaunch period. There is no substitute for good personnel.